WHAT WOULD WALT DO?

Life Lessons Learned When Building Walt Disney World That Still Apply Today

Second Edition, 2017

TABLE OF CONTENTS

INTRODUCTION

Most of us at one time or another have said, "I could write a book about that!" This is a little book I finally got around to writing. Much of this book's contents were first published as a soft cover by iUniverse under the same title in 2001 with ISBN: 0-595-17203-2. This 2017 version has been abbreviated from the first and includes photographs which the first version did not.

I learned many life lessons about quality and integrity from working on the construction of Walt Disney World in Orlando.

In the 1960s I received a Civil Engineering degree from the University of Florida. Circumstances led me to the right place at the right time. Less than two years after graduation, I found myself working on the construction of Walt Disney World near Orlando, Florida. I managed a firm that provided quality control inspection and testing services on most of the construction. That perspective allowed me to see everything get built from the ground up.

Like most people, I was a fan of Walt Disney and had enjoyed his movies and television productions all of my life. Walt's love of people and knowledge of how to entertain them was the hallmark of his career. He knew what we liked, and was never afraid to risk everything to bring it to us.

Walt died in December 1966, more than a year before construction began on Walt Disney World. Many of the people who helped build Walt Disney World had worked for Walt on projects when he was still alive. When faced with problems during design and construction, they would always ask each other the question: "What would Walt do?". The quality of the Magic Kingdom and the rest of Walt Disney World proves that they knew what Walt would have done. They did it Walt's way.

One of Walt's dreams was what he first called a "City of Tomorrow" and which he eventually named Epcot, an acronym for Experimental Prototype Community of Tomorrow. He wanted Walt Disney World to produce the money that would allow him to build his dream city. Epcot would be a working city with homes, businesses, and a showcase for the latest technology of American industry. It would have no slums, no poverty, no crime, unlike the major cities of the real world in the 1960s and even today.

He was not around in person to make sure the Epcot of his dreams came into being. The Epcot of today is a wonderful attraction, but it does not accomplish those things that Walt wanted to achieve in his futuristic city.

This small book tells the story of some of the people who helped build Walt Disney World and how Walt's spirit guided them and generated the question "What Would Walt Do?

CHAPTER 1. THE WALT DISNEY LEGACY

Construction of Walt Disney World began shortly after Walt Disney died in December 1966. Even though Walt was not there in person, it seemed like his spirit was very much present and involved in almost every key decision.

Any construction project, especially a large one, has to be managed carefully to balance three constraints: time, cost, and quality.

TIME: *the immovable opening date for Walt Disney World was October 1971.*

COST: according to Roy Disney, *the budget for the project was $125 million.*

QUALITY: *the theme park had to be built as closely as possible to Walt's ideas and with the guest's experience and enjoyment foremost in mind.*

It's easy to see that each constraint affects the other: increase the quality, for example, and you increase the cost and time. Since quality and schedule were inflexible, the cost of the project exceeded the budget due to overtime work required. **The actual cost was $400 million, which is more than $2.5 billion in 2017 dollars.**

The following imaginary - though typical - meeting demonstrates the issue.

Ω

"What would Walt do?"

The question echoed in the stillness of the conference room. The room was in a construction trailer on the muddy site of Walt Disney World near Orlando, Florida. Seated around the table were an engineer, an architect, a construction manager, a contractor, a Disney art director, and a Disney project manager. Blueprints were spread on the table. Cigarette smoke filled the small room.

It was early 1971, and the opening of Walt Disney World was scheduled for October of that year. It had to open in October. Package tours had been sold and television specials committed to. It had to open in October, no matter what. There was no choice.

Schedules were slipping and there was pressure to relax some of the strict standards that had guided construction up to then. The conversation at the table had been heated.

"The exterior walls on the Main Street Railroad Station are not the texture we wanted," the art director complained.

"But it doesn't look bad to me, "the engineer replied. "I think it's good enough. No one will ever notice the difference. If we try to change it now, the schedule could slip even more. We have to make that October opening."

"The surface needs to be sandblasted off and redone as originally designed," said the architect.

"It just doesn't look authentic. It doesn't fit the period it is supposed to represent. The contractor screwed it up and needs to make it right," said the art director.

The contractor snuffed his cigarette out and leaned back in his chair.

"I didn't screw up", he said. "Your plans were not clear on what the texture needed to be. If you want to change it now, it will take additional money to get it done. And if you want it done without causing schedule delay, you will also have to pay extra for my men to work overtime. Think about it. Is it worth it to you?

After a long silence the Disney project manager, in a quiet voice, asked no one in particular, **"What Would Walt do"?**

Walt Disney, the ultimate art director, the genius whose spirit still filled the construction site more than four years after his death. After a few minutes of deliberation, the project manager answered his own question.

He turned toward he contractor.

"Sandblast the walls and do it the way we originally wanted. My art director will work with your people to get the exact surface we want. We will pay you for the extra work and the overtime. Let's just get it done the right way, the way Walt would have done it," the project manager said.

The Question That Defined Walt Disney World

Walt Disney died in 1966, and that question – What would Walt do? - was asked hundreds of times between 1967 and 1971 by the designers and builders of Walt Disney World in Orlando, Florida. It was asked about financial decisions, quality considerations, creative choices, and employee relations.

The question was asked whenever a tough problem or obstacle popped up. The answers always came quickly and usually seemed obvious to the people who had worked with Walt on his movie projects and on Disneyland in Anaheim, California.

The answer to this question guided everyone involved in the creation and construction of Walt Disney World.

CHAPTER 2. WALT'S EARLY DAYS

Walter Elias Disney was born in Chicago, Illinois, on December 5, 1901. When he was only four years old, his parents decided to move the family from Chicago to get away from the crime and squalor that were typical of large industrial cities at the turn of the century.

The family moved to Marceline, Missouri, a small town about 120 miles northeast of Kansas City. Walt's parents, Elias and Flora, had five children: Herb, Ray, Roy, Walt, and Ruth. Walt was especially attached to his brother Roy, who was eight years older. They would enjoy a close lifetime personal and business relationship.

Although the family lived here for only a short time, Walt always considered it his hometown. His strongest memories were of the people and places and events in this small town. Years later, when he was weatlhy and famous, he would call people to the train window when they were going through the countryside around Marceline to show off his hometown.

Walt Volunteers to Serve in World War One

Walt enjoyed artistic success at a young age. When he was still a boy he began drawing and painting pictures that he sold to his friends and neighbors. After dropping out of high school and lying about his age, he served in World War One as a volunteer ambulance driver for the Red Cross. He was still a teenager when he came back from Europe and began to work for an advertising agency. When he was only 20 years old, he started his own small studio, Laugh-O-Gram, in Kansas City, Missouri. The studio was financed by money from small local investors.

Walt Starts Laugh-O-Gram and Fails at Age 21

Disney created and drew some notable cartoon characters at Laugh-O-Gram and produced a cartoon called *Alice's Wonderland* which featured a real little girl interacting with animated cartoon characters. The film got good reviews and was very popular.

Financial success was more elusive, however, and when Walt was only 21 years old, he had to declare bankruptcy and fold Laugh-O-Gram. It embarrassed Walt, but he worked hard at paying of his debts. Eventually, all of his creditors would get 45 cents on the dollar. He wanted to leave the scene of his financial embarrassment and decided to move to Hollywood and be a director. This was in the early days, movies were still silent, and animated cartoons were pretty primitive. The problem was Walt didn't have any money to get out to California.

He solved the problem by raising the cash for the trip by taking pictures of babies. He raised additional money by selling his camera when he finished the baby picture project. He bought a first class train ticket to Los Angeles. The first class ticket was a symbol to Walt of nothing but the best. He was broke after buying the ticket, but he traveled in style.

From the beginning of Walt's career, his partner was always his brother, Roy. Walt never worried about money and Roy always did. In the early days of the Disney Brothers Studio, Walt and Roy lived together in a small apartment and shared cooking and housekeeping duties. They had total trust in each other and although Walt was the boss he relied on Roy to figure out how to finance the projects.

Walt Has a Hit with Steamboat Willie

Walt's success in the entertainment industry is well documented from his hit with *Steamboat Willie*, the first cartoon featuring Mickey Mouse to later box office hits of the 1930s and 1940s like *Snow White, Pinocchio*, and *Bambi*. For years, Walt was Mickey's voice on these cartoons, and his animators began to study Walt's mannerisms and took photos of him. Mickey Mouse became more and more like Walt over the years. He always identified strongly with Mickey.

In 1934 he decided to do a feature length animated movie. He risked every penny he had on this venture when almost everyone in his personal life and in the industry told him it would be a financial disaster. He knew what he wanted and he believed in himself. It would not be the last time he risked everything to follow his own dream.

Walt Makes Snow White and Hits the Big Time

The feature length animated movie was *Snow White* and Walt personally gave each of the Seven Dwarves their names. *Snow White* was a big hit, and the brothers were able to pay off all their debts for the first time. They bought their parents a bungalow in California and continued to crank out animated features.

Walt had begun to dream of building an amusement park back in the 1940s. He wanted to create a park for families where parents could take their children and where both adults and kids would have fun. He was turned off by the amusement parks of that day, considering most of them to be cheap and sleazy. He thought he could build a different kind of park, one that would never decline into the ugliness that many of the leading amusement parks in the country had fallen into.

The Disney planners worked on ideas for this park at the Disney Studio in Burbank but soon realized that the small lot available for the park was not nearly large enough to contain Walt's ideas. In general, his friends, and especially his brother Roy, did not take this dream too seriously. The company was struggling with cash flow problems and was in the middle of several movie projects, so Walt tucked his park idea away in the back of his mind for better times.

The military took over The Disney Studios during World War Two, and Walt produced many training films for the U.S. Government. These films apparently cost the Disney Studios more to make than they received, and many of his top assistants were lost to military service. It was years before the company fully recovered from the impact of the war.

By the early 1950s the studio was back in full swing, making money, and had done their first live action feature film, *Treasure Island*. The company had grown to hundreds of employees. Although Walt always tried to treat his employees well, he could no longer know each of them intimately on a personal basis. There were just too many of them.

The union came on the scene and organized many of Disney's workers into two separate unions. One of the unions went on strike and some of the picketers would shout insulting things to Walt when he came to work. He never got over the strike. He felt betrayed by his own friends and workers. Before this strike, he had always been very socially liberal. After the strike he became very politically conservative. He believed that the unions were heavily infiltrated by Communists.

After the dust settled from the union turmoil, Walt began to think about the theme park idea again. He was calling it Mickey Mouse Park in those days. He had many sketches done and was starting to formulate how he wanted it laid out and what its character would be. The name Disneyland was starting to be used.

Walt Buys a California Orange Grove and Plans to Build Disneyland

The 11 acre parcel across the street from the Burbank studio was too small to contain Walt's ideas. Walt hired consultants to help him find property in the ideal location. These consultants settled on Anaheim for a number of reasons. Walt purchased a 160 acre orange grove in Anaheim, and work began in earnest to complete plans for Disneyland.

Walt figured out a neat way to finance the new park. Television was a new phenomenon back in the early 1950s. Many families were just discovering the entertainment value of the little white screen and wanted a TV set for their own homes. Walt figured the television industry needed family-oriented programming. He approached the networks with his ideas. He was rejected by NBC, but ABC was interested in hearing more about Walt's ideas. What resulted was a one year long television program on ABC that began on October 27, 1954. The show gave viewers periodic updates on the construction of Disneyland. The rest of the show was filled up with Disney features like *Davy Crockett*. The song, *"The Ballad of Davy Crocket"* became number one on the Hit Parade for 13 weeks. The program was a tremendous financial success, and made big stars out of Fess Parker and Buddy Ebsen.

Walt and his teammates knew that Disneyland would become a reality. Construction began on the park in August, 1954. Walt announced that Disneyland would open to the public in July 1955, only 11 months after the start of construction. It was truly an ambitious schedule.

CHAPTER 3.   DISNEYLAND IN CALIFORNIA; WORLD'S FAIR

Construction began on Disneyland in August, 1954, a couple of months before the *Disneyland* show began its run on ABC. The profits from the Disney television show continued to come in and the new park was scheduled to open in July, 1955, a construction schedule of only eleven months. Walt's consultants met with experienced amusement park operators who predicted the park would flop. They said it didn't have enough ride capacity, had only one entrance, and listed many other reasons why it would fail.

Walt Hires Admiral Joe Fowler to Build Disneyland

As always, Walt ignored the doomsayers and followed his own instincts. He hired retired Navy Admiral Joe Fowler to build his park.   Admiral Fowler was a master big-time construction project manager who had supervised the construction of many shipyards and other naval facilities during World War Two.

The project was plagued by union strikes and the confusion of Orange County, California, building inspectors.   The inspectors had never seen anything like Cinderella's Castle or The Swiss Family Tree House or the other innovative structures. A lot of time had to be spent convincing them that the buildings were safe. Another problem was getting men and material out to the Disneyland site. It was a long way from Anaheim to anywhere else in southern California back in 1954.

Admiral Fowler was the type who set out to do something and let nothing stop him once he had started. In spite of these problems which might have defeated a lesser project manager, the Admiral figured out ways to forge ahead and get the job done.

Walt was personally involved during the entire project. He had directed the planning and layout, and had a layout prepared by Herb Ryman that was used in lining up the original financing and community support. Walt's personal touch was in the design of Sleeping Beauty's Castle, Jungle Cruise, Haunted Mansion, Pirates of the Caribbean, Rivers of America and most of the other attractions at Disneyland.

At the beginning of the project, Walt could hardly read a blueprint. By the time Disneyland was finished he was an expert at reading plans. Not only that, he knew everything about every system in the park. Walt knew where every pipe was, what it did, how all systems worked, and how high every building was.

An example of Walt's belief in his own judgment involves the Swiss Family Tree House. The tree was made of massive steel plates welded together to make a huge trunk, and then "branches" made of steel pipe were attached to the trunk. The whole artificial thing was then covered with fiberglass to look like tree bark. Artificial leaves were then attached to the branches. It was extremely expensive construction.

Walt came to the site one day during construction when the tree was more than half finished. He walked around the tree several times, looking at it from different angles, casting a critical art director's eye over the tree and its relationship to surrounding features. He called in Admiral Fowler and told him the tree had to be moved to a new location a few dozen feet away. The admiral pointed out how expensive this would be and how it could impact the opening date. Walt told him it didn't matter. The tree didn't look right where it was. It had to be moved regardless of the expense but the opening date could not be delayed. And so it happened.

Disneyland opened on July 17, 1955, on schedule in spite of all of the problems involved in creating a brand new kind of entertainment park. Admiral Fowler said that Walt used to come to Disneyland after it opened and just sit on a bench and relax and people watch. He told Joe that it was his escape from the stress of making films. He said his home and Disneyland were the only two places where he could truly relax.

Both Walt and Joe Fowler were involved on a daily basis with the operation of Disneyland until Walt's death. One of Walt's special concerns was how guests and staff would relate to one another. He insisted that all Disney employees be on a first name basis. They were instructed to call him and all supervisors and managers by their first names.

One day he walked into a Disneyland restaurant for lunch. A waitress came to his table and asked, "Can I help you, Mr. Disney?"

"Yes," he replied. "But call me Walt. The only Mister at Disneyland is Mr. Toad."

Once Disneyland opened, its success was obvious and the area around Anaheim exploded with development. Residential neighborhoods sprang up almost overnight and so did tall hotels, restaurants, tee shirt shops, souvenir stands and other businesses selling things to tourists. Walt was disgusted with the way all of this looked but was unable to control it. The best he could do was to make Disneyland an oasis of calm and good taste smack in the middle of urban blight.

Walt Puts Four Attractions in the New York World's Fair in 1964

Disneyland was a tremendous success but Walt was not completely happy with it. He began to think about developing another park somewhere with plenty of land so he could control what went in around it. He knew the new park would have to be far enough away from Disneyland not to be a competitor. He also knew it had to be much larger than Disneyland's 160 acres so that in addition to the buffer there could be plenty of room for new ideas and new attractions.

Disney tested the waters for his new ideas by putting four attractions in the New York World's Fair in 1964. Walt was again involved very heavily in all of the ideas and designs. He planned to move these attractions back to Disneyland after the fair was over. He also wanted to see if the people in the northeast would like his attractions.

Walt Hires General Joe Potter

Walt met **General William E. "Joe" Potter**, a retired Army Corps of Engineers officer on this project. General Potter was the Executive Vice President of the World's Fair and had been governor of the Panama Canal Zone during his military career.

The fair was a big hit with the New Yorkers and others that attended. The Disney attractions were Primeval World; It's a Small World; Progressland, and Great Moments with Mr. Lincoln.

The World's Fair taught Walt and his team many lessons that would be useful in years to come. They learned new ways of moving large crowds of people through their exhibits. They also learned a lot about audioanimatronics, a term they invented to describe the machinery and controls that moved Abraham Lincoln and some of the other characters in the fair. Walt had also been impressed with General Potter's capabilities and put him on the Disney payroll back in California.

CHAPTER 4. WALT HAS A BETTER IDEA FOR AMERICAN CITIES

The success of the New York World's fair gave Walt the confidence to begin putting together the property for a new theme park which the Disney people called Project X. The movie "Mary Poppins" was making a lot of money for Disney and it came just in time to produce the dollars that would be needed to buy the acreage.

Walt founded a separate company, MAPO, which was short for Mary Poppins. MAPO carried on the duties of developing the Florida project for a while until it was absorbed by WED Engineering and the Imagineers. WED stood for Walter Elias Disney and Imagineer was a Disney word combining "imagination" and "engineer". It is one of those perfect words that is now in many dictionaries.

The City of Tomorrow Becomes Walt's Mission in Life

Walt and General Potter began to work on concepts for the Florida project. Walt's main focus was the overall concept of the Florida Project and especially what he originally named the **City of Tomorrow**. It was assumed the theme park would be pretty much like Disneyland, only bigger. The City of Tomorrow, however, became Walt's mission in life and he spent incredible time and energy pursuing it.

Disney sent letters to hundreds of corporations. He wanted to find out what industries were looking to do in the future. He wanted to discover what was working in the new towns in the world and what was not working. He studied city planning and hired consultants to do studies on planned cities.

Walt and General Potter visited hundreds of research centers, factories, and new towns during this period of investigation and research. The Disney plane criss-crossed the country with Walt Disney and Joe Potter in search of the best ideas. A story circulated about Walt on some of these flights that demonstrates why he had so much empathy with the common man.

The Disney plane that Walt used on most of these trips was named **Mickey Mouse One**. It was a Grumman Gulfstream prop-jet that carried Walt and other Disney executives on company business. A typical flight might find the plane somewhere over the heartland of America when Walt would amble up to the cockpit and start chatting with the pilot.

"Where are we now?" Walt asked.

"Close to Tulsa, Oklahoma," the pilot replied.

"Do they have one down there?" Walt asked.

The pilot checked a tattered paperback directory in the cockpit.

"Yep, they sure do, Walt. They have a couple of them."

" Well, lets go on down then," said Walt.
So the plane would land at the Tulsa airport. A taxi would take Walt and his party to the nearest McDonald's restaurant. You see, Walt was having a Big Mac attack. He loved McDonald's, both the food and the family image that the restaurant chain had perfected.

As a result of his research, Walt believed that most of the hundred or so new towns in America were failures. He wanted to plan communities where people could reach their full potential in a world becoming more hectic by the day. He believed that some of the model cities had been doomed to failure by old and unimaginative building codes, labor unions that were only interested in protecting their jobs, incompetent and argumentative building contractors, and selfish and narrow-minded politicians.

Walt vowed that these failure modes would not be part of his City of Tomorrow. He felt the city could be a research and development laboratory where American companies could test new concepts in housing, transportation, and manufacturing. They could test their concepts without being hampered by old fashioned building codes, restrictive union labor practices, and the general negative bureaucratic thinking which he felt was typical of most American cities, counties, and States.

Walt's original concept was of a domed city. Since his original concept, domed technology has advanced tremendously, as demonstrated by all of the new major stadiums in the country. Back then, however, his idea was extremely innovative and daring. People would live and work under an environmentally controlled glass or Plexiglas umbrella. There would be no pollution; only the purest air and water would be contained within the transparent sphere.

One day Walt thought it was time to give a name to his future city. He decided on **Epcot**, which stands for **Experimental Prototype Community of Tomorrow**. Many of the people who knew Walt felt that he would not have done Walt Disney World at all if it wasn't a way for him to achieve his new city. He felt he could help solve some of the problems of urban life which were plaguing America. He believed that government could not make the big changes required, that American business and industry were the key to the success of any new town.

With the idea for the new city burning away in his mind, Walt began the search for a location that could accommodate his dream.

CHAPTER 5. ORLANDO IN THE CROSSHAIRS

Walt and his team began looking for locations for another Disneyland in the early 1960s. They eventually focused on Florida and looked at parcels near Ocala, West Palm Beach and Daytona but a lot of factors led them to the Orlando area. Walt did not want his new Florida Disneyland to be too near the ocean. He didn't want the competition of the beach, plus he was aware of the fury of hurricanes and the damage one could inflict on a theme park.

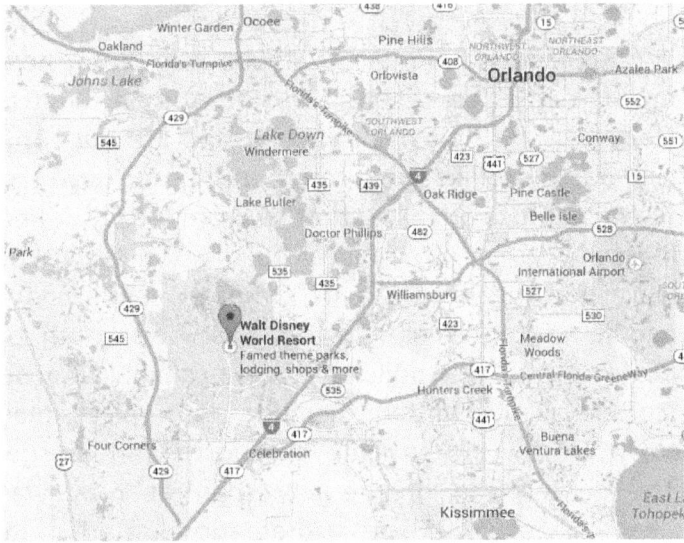

Another important factor was that Orlando and Central Florida were within a one or two day drive for millions of potential customers east of the Mississippi River. This market was one that Disneyland way out in California never fully tapped. Orlando was at the intersection of the Florida

Turnpike and Interstate Highway 4. The Turnpike funneled people from the Midwest and South into the Florida peninsula and I-4 from Daytona Beach carried folks from the East Coast of the country who had come down I-95.

Another big factor in the decision was the low price of land in Central Florida in the 1960s. Walt wanted lots of land to surround his new creation. He hated the squalor and commercialism that surrounded his small park in Anaheim. He referred to the area around Disneyland as a "cheap Las Vegas."

The City of Orlando itself was an example of how much lower real estate prices were in Central Florida at the time Walt was looking around. The city was more like a big town; it did not have the high rise office buildings that so dramatically define its skyline today. The tallest buildings in the 1960s were the Angebilt Hotel on Orange Avenue and a couple of retirement towers east and south of downtown. These buildings were probably 10 or so stories tall. Downtown Orlando had typical small town Florida stores like Woolworth and Kresge. The centerpiece of the city then as now was beautiful Lake Eola with a dramatic fountain in its center and swan boats floating serenely around the lake.

Walt and his team began to focus on an area about 15 miles southwest of Orlando. The land they were looking at was north and south of U. S. Highway 192, and along I-4. Some of it was in Orange

County, the rest in Osceola County. The location was far enough away from Orlando to be inexpensive, but close enough for easy access from Orlando, both during construction and later when the park was up and running. The area had no real name back then, but most of it was near a tiny village known as **Vineland**.

Vineland was on the eastern edge of the Disney property. It was just a place name on the map less than a half mile north of the intersection of State Road 535 and Winter Garden-Vineland Road. In those pre-Disney days, it seems like Vineland was the center of that part of Orange County. Even today the roads ring out this truth: Apopka-Vineland Road, Winter Garden-Vineland Road, Taft-Vineland Road and Orlando-Vineland Road. Many of these roads had originally been the routes of small railroads that served the citrus industry in the late 1800s and early 1900s.

Rumors began to circulate around central Florida in 1964 and 1965 that somebody was quietly assembling acreage northwest of Kissimmee, a small cattle town south of Orlando. Local people speculated that since Orlando was so close to Cape Kennedy, the buyer might be an aerospace firm like Lockheed or McDonnell. Other rumors suggested that one of the big automakers like General Motors or Ford was going to build a plant out there. A handful of people speculated that it might be Disney trying to put together a big parcel for a Disneyland East. If anyone knew for sure, they were not talking about it.

The secrecy involved is still a Central Florida legend. If key people in the area like realtors, developers, citrus barons, and ranchers knew who was behind the purchases, they kept it quiet. Disney set up hundreds of corporations to buy the land. By concealing the real buyer, they were able to keep prices down and prevent anyone from figuring out that it was only one buyer. The Orlando Sentinel apparently figured out what was going on, but Martin Anderson – owner and publisher of the Sentinel - kept it quiet at Walt Disney's request.

**Walt Disney, Governor Hayden Burns, Martin Anderson
In November 1965 – State Arichives of Florida**

The largest single purchase Disney made was ranch land of more than 10,000 acres from a pioneer Osceola County family, the Bronsons. The family was headed back then by **Senator Irlo Overstreet Bronson**, a Florida state legislator and prominent cattleman. The Bronsons did not know who the true buyer was.

Senator Irlo Overstreet Bronson, Sr.
State Archives of Florida,

When the Bronsons learned who the buyer was, some advisors told the senator he could get out of the deal and hold out for more money. Bronson refused, saying that a "deal is a deal". Disney also bought hundreds of little lots owned by people who had never seen their land. These were lots that had been purchased sight unseen by hopeful dreamers long ago during the Florida land boom of the 1920s.

**Oren Overstreet Brown on the right
talking to customers at Brown's Cafe**

Not everyone sold their land to Disney. One man became famous as the man who said no to Disney. His name was **Oren Overstreet Brown**. He reportedly turned down an offer of $4.2 million for his 6,750 acres of ranch land and Reedy Creek swamp abutting what would later become Walt

Disney World. Brown was from a pioneer Osceola County family, the Overstreets. He owned and operated Brown's Café and pool hall in downtown Kissimmee and also served seven terms from 1957 to 1985 as an Osceola County Commissioner.

Brown had no particular problem with Disney, but just preferred to keep his land. In a Look Magazine article in 1971 he was quoted as saying,

"What's money? It's only paper, most of it....I never could keep money. The land, it won't run off. Lots of people like money, but I don't care much for it. I reckon I'm peculiar that way."

Walt Disney, General "Joe" Potter, Roy Disney and Governor Hayden Burns - Orlando, November 15, 1965. Florida State Archives

Disney had to settle for less land than they wanted and finally bought about 27,000 acres at a total price of $5 million through all of the corporations they had set up. The average price was about $185 per acre.

Many local lawyers and real estate brokers made small fortunes helping Disney pull the properties together. Not long after that regular people in the area would start making good money too working on the construction of the project and working at Walt Disney World when it opened.

With the real estate safely in hand, it was time to make the formal announcement to the world that Disney was the buyer and had big plans for the property. A press conference was held in Orlando in November, 1965. Walt and Roy Disney were there along with the movers and shakers in local and state politics.

It would have been hard for anyone at that press conference to imagine that Walt would be dead in just a year. He was alive and happy and animated as he discussed plans for the property.

Walt began to finalize the plans for Florida Disney World, as it was then called within the company, but his health was beginning to fail. For years, his trademark around the studio had been his cigarette cough. Animators could tell where he was at any time by the cough announcing his presence. Finally, the doctors discovered lung cancer and had to remove a lung in November 1966.

His health failed rapidly after the operation, and he died on December 15, 1966. But the dream of Walt Disney World and EPCOT had been launched. Walt had surrounded himself with people who would help build his dream, people like Admiral Fowler and General Potter.

The spirit of Walt would be very much alive during the design and construction of Walt Disney World. That spirit would capture the interest and passion of thousands of people Walt had never met who would help him build his dream.

CHAPTER 6. THE EARLY DAYS OF CONSTRUCTION

By early 1968, the whole world knew what Disney was planning in Orlando, but it still seemed unreal, especially when you looked at how small and unprepared for the big event Orlando was.

The Orlando airport in those days was really McCoy Air Force Base. Several military buildings and hangars squatted on the west side of the runway. A small concrete block building on the east side of the runway was the Orlando terminal. It was similar to other small town Florida terminals. The airport managers rushed to build a new terminal to handle the anticipated crowds that Disney would generate.

McCoy Air Force Base, Orlando, Florida 1960s

An attractive new terminal was completed around the same time Walt Disney World opened. This building is still visible from the Beachline Expressway just west of Semoran Boulevard (so named because it connected SEMinole County to ORANge County). The new building was too small the day it opened, and the airport soon expanded and relocated its terminals to where they are today.

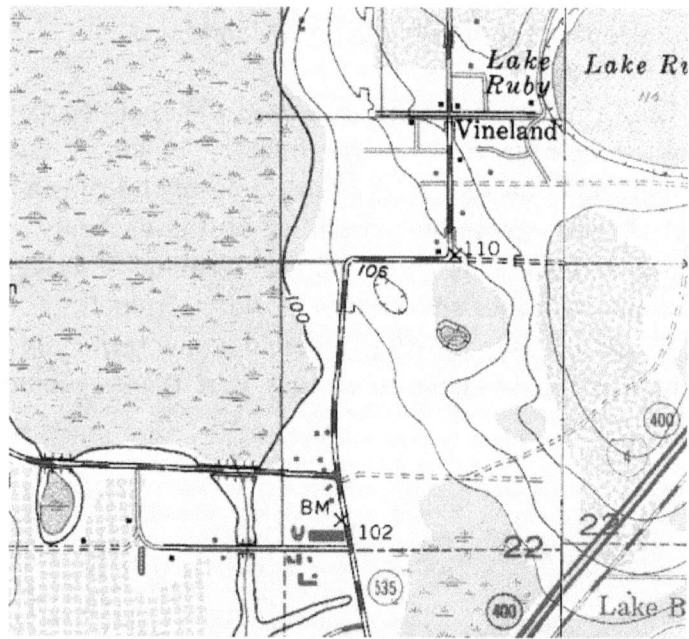

U.S. Geological Survey Map of Vineland, Florida Pre-Disney

State Road 535 passed under I-4 about 17 miles west of Orlando. The only structure at that intersection back in 1968 was a Stuckey's Restaurant. It sat at the northeast corner of the interchange. Less than a mile from where it passed under I-4, State Road 535 took a hard left and became known as Winter Garden – Vineland Road. If you continued straight instead of turning on 535, the road was known as Apopka-Vineland Road, and it ran north through the sleepy place known as Dr. Phillips, where nothing more than an old citrus packing house existed by the side of the railroad tracks which headed south toward Vineland.

Gee and Jensen of West Palm Beach was the engineering firm in charge of designing and constructing the massive drainage system within the Reedy Creek Improvement District, which included the Magic Kingdom. Gee and Jensen was founded in West Palm Beach by Colonel Herbert Gee, formerly of the U. S. Army Corps of Engineers.

Roy Disney, General Joe Potter, and Card Walker on Bay Lake in 1967 - State Archives of Florida

I always assumed that General Potter must have had something to do with hiring an old Army engineer to be in charge of this work.

The project had many miles of canals, and in keeping with Walt's desire for a natural look, the canals were curvilinear rather than straight. Flooding of the flat piney woods and swamps was a potential problem. To mitigate the flooding problem, retention areas and wetlands along the canal network had their water elevations regulated by an automatic water level control gate called an AMIL. This was a huge metal gate that hung like a moveable dam across a canal. When the water level on one side rose to a certain height above the water level on the other side, the gate would gently open and allow the water to go from high to low. It required no power, and was a quiet and efficient operation.

Roy Disney, General Potter and other Disney executives were familiar figures as plans were being finalized for projects on Bay Lake and other natural areas on the site.

Bay Lake was a natural body of water just to the east of what would become the Magic Kingdom. Roy Disney did not like the murky brown color of the natural water. The water was tinted by natural vegetation such as cypress, bay trees and moss. It was the color of ice tea. Roy asked Gee and Jensen to come up with a way to clarify the water.

Hundreds of test borings had been drilled all over the Disney property, and it was discovered that the murky waters of Bay Lake were underlain by pure white sand. The mucky bottom soils were removed by dredged and relocated around the site as landscaping soil. Then the pure white sand was dredged up and used to create nice white sand beaches around the lake.

A canal was dug around the lake to direct the natural runoff away from the lake. An AMIL gate was installed on the discharge canals so that the lake would have a constant water level.

Aerial View of Bay Lake in June, 1967, State Archives of Florida

A favorite spot for a sausage sandwich breakfast in those pre-heart healthy days was that lonely Stuckey's at State Road 535 and Interstate-4. The manager of the store was an elderly man with a bald crown and dark rimmed glasses. He looked somewhat like Sergeant Bilko as played by Phil Silvers on the old TV show.

He had invested most of his life savings in a vacant lot at the off ramp across from Stuckey's on State Road 535. He sold it a few years later to an oil company for a fortune and was able to retire. A Chevron gas station is on that property now, but the Stuckey's is long gone, replaced by a modern business building.

Most of the Disney property was swamp in those days. The process of getting test borings and soils samples involved dragging a johnboat and a tripod drill rig for miles through the swamps. The swamps were populated with water moccasins, gators and bloodsucking leeches. Gee and Jenson had an office in Kissimmee set up to work on the project, and **Walt Stephens** was in charge. He directed the work on test borings for the drainage canals and structures.

The Disney staff was in a single story small building not far from Stuckey's. It was located near Bay Lake not far from the current location of Disney's campground, Fort Wilderness. **Leonard Wood** was Disney's surveyor on the property. He worked for VTN, a California company with Disney experience from Disneyland. Leonard drove a Toyota Land Cruiser, which was an unusual vehicle in Florida back then. Japanese cars were far more common in California than in the East.

Signs were painted on the office doors in the small Disney building. A brassy military flavor showed through with the names and titles: Admiral Joe Fowler, General William Potter, Commander Ray

Fail. The enlisted types always got a kick out of the door that was labeled Sergeant Leonard Wood.

Mornings at Disney World During the Construction Days

Many people who are not naturally morning persons become one by necessity if they are working on a big construction project. Most workers got to the Walt Disney World construction sites around 630 in the morning.

They learned that early mornings in Florida can be a good time to get in touch with their senses and get their thoughts in order. Ben Franklin was probably on to something when he noted that early to bed and early to rise makes a man healthy wealthy and wise.

The Central Florida roads leading into the Disney construction site were often foggy in the early mornings. As the skies began to lighten in the east with the coming sunrise, the fog took on an eerie dim glow with the brighter halos of automobile headlights moving back and forth in the semi-darkness. It made travel to work a nervous adventure.

In the springtime the smell of orange blossoms filled the air during the morning drive and brought back early memories of a Florida where the blossoms bloomed all over the state, north and south. As the sun finally peeked over the horizon, dozens of hungry construction workers pulled into

Stuckey's at I-4 and Apopka-Vineland Road for a cup of coffee and sausage sandwich. Others would stop at Johnny's Corner for their small breakfast snack offerings.

Early mornings on a big construction site are a special time. Workers are shaking out their tools and a few machines are beginning to rumble and whine and clatter. Backhoes are dipping into the earth and look like grazing dinosaurs in the dim early morning light.

The early morning central Florida winter can be nippy and even freezing. Fires are lit around the site in open drums and in oil fired chimney type warmers called salamanders. Construction guys are standing around warming up their hands, people are saying hello to each other, and the new day is beginning.

Vapor drifts into the air from puddles on muddy areas of the site. Workers breathe out wisps of steam mixed with the ever-present cigarette smoke. The few birds that haven't been scared away from the treed fringes of the site begin to twirp and tweet mysterious feathered messages. A new day on a construction site is even brighter than in most of the rest of life. Workers are renewed and ready to try again. They have a goal in sight: get this thing finished. They have a purpose in being.

A Cold Wet Morning in Reedy Creek

Walt Disney World was built in and among the swamps of Central Florida. It was difficult country to get around in when test borings were needed or areas had to be inspected. Sometimes a four wheeled or tracked vehicle could not do the job. Another approach to the problem was considered.

Amphibious Vehicle Known as Coot From 1965 Brochure

One cold January morning a salesman came out to demonstrate a four-wheel drive amphibious vehicle known as the Coot. It had big heavily treaded tires, and you could also hook up a propeller to a shaft that came out of the engine on its rear end. . It would float and go through the water just like a boat. The Coot was designed to be able to fit into the bed of a standard pickup truck.

The salesman offloaded the Coot on the bank of Reedy Creek near where it crosses U.S. Highway 17-92 in the wilderness on the south side of the Disney property. Two Law Engineering guys and a Disney surveyor joined the salesman. All four men climbed into the Coot, two in front and two in back. The Coot chugged along in grand style into the woods on the soggy ground.

Reedy Creek On A Cold Quiet Winter Morning

The little vehicle was great as it plowed along, never bogging down in the sticky muck and gumbo clay. The exhaust pipe from the sturdy little engine puffed out wispy strands of white fog into the cold Florida morning air.

The salesman wanted to demonstrate how the Coot performed in the water, so he stopped at the edge of the creek and clamped the propeller onto the shaft at the rear of the vehicle. He climbed back in and steered the Coot into the deep waters of Reedy Creek. It bobbed along in the cold water and for a few seconds everything seemed fine. The Coot traveled down the middle of the stream at a good clip. Then suddenly it began to list to one side and water began pouring over the edge of the vehicle. In a split second, the Coot sank like a rock.

All four men suddenly found themselves in 7 feet of very cold creek water. When they got over their shock, they began to dive down in unison and dragged the Coot into shallower water a painful few inches at a time. At last they got it into shallow water and were able bail out some water with their bare hands. They slowly kept dragging it closer to shore. It was so cold their teeth were chattering and their lips were turning blue.

After what seemed like hours, they finally got the vehicle up on the bank. The salesman removed the spark plugs and cranked the engine by hand. Water spurted out until the engine finally seemed to be fairly dry. He put the plugs back in, cranked it, and the little engine sputtered to life. The frozen adventurers made a hasty escape from the cold swamp, carefully avoiding deep water.

This incident underscores the importance of a salesman understanding his product. The manual said the capacity of the Coot was four people on land, three in the water. The salesman didn't read the manual. The weight of the fourth man was enough to sink the vehicle deeper in the creek so that the water flowed in.

All three of the salesman's guests denied being that fourth man.

Johnny's Corner in 1957. Apopka-Vineland Road in the center, Winter Garden-Vineland Road heading left

The State Road 535 Vineland exit on I-4 was a desolate crossroads when work was beginning on Walt Disney World. That exit today takes the traveler into a bee hive of tourist activity, including hotels, restaurants, gift shops, shopping centers, and gas stations. The modern tourist would have a hard time believing that all that existed back in 1968 at that intersection was a small concrete block house that had become a beer joint and country store known as **Johnny's Corner**.

During the construction of Walt Disney World, Johnny's Corner was the nearest place to cash a paycheck and buy some beer. They reportedly sold more beer than any other retail outlet in Florida – maybe even in the entire United States.

Law Engineering's Office Trailer Gets Burglarized

One morning I came to work and my office and lab trailer complex was surrounded by Orange County Sheriff's Department cars and Disney security vehicles, red and blue lights flashing away. I went into the office trailer and learned that someone had stolen all of our typewriters and our petty cash box. The deputies were dusting the place for fingerprints. I didn't know how that would help, because all they would find would be the prints of my employees. It could have been anyone. Plus a lot of our clients came around the place, so their prints would also be there.

A few days later the sheriff's department called and asked me if any of my workers had failed to come to work recently. I told them that Jim Smith was the only one. They asked me for Jim's social security number, and called me back later in the day to say it was a fake number. Jim Smith was really Bob Wilson, and his prints were all over the place. They told me then that he was an escaped prisoner from Illinois. We were so hard up for good help back then, we didn't do a good job checking references. Jim or Bob, whoever, was a good worker who never made any trouble. He would be a hard guy to replace.

Johnny's Corner

In 1969 the only building at the corner of State Road 535 and Winter-Garden Road was an old concrete block house which resembled the thousands of others which had been built in Florida in the 1950's and 1960's. The modest single story house had been converted into a big open bar room and was known as Johnny's Corner. This humble structure was packed during weekday afternoons after work.

Bob, Johnny's Corner Co-Owner, Serving a Construction Worker, Circa 1970

Johnny's was the place where construction workers would let off steam after work. The southeast corner of that intersection, where it branches off left to Winter Garden-Vineland Road, was the main route at the time for all construction traffic into the project. Thousands of construction workers drove by twice a day. Johnny's was owned by a couple of regular guys, neither of them named Johnny. One was named Bob. They saw an opportunity to sell a lot of beer and worked hard doing it. They ran a pretty good place.

A national magazine ran an article on Johnny's Corner back during the hey-days of construction. It was reported that Johnny's sold more beer than any other retail outlet in America. It was not necessarily a boom town saloon loaded with drunken workers spoiling for a fight, but it definitely was a place where a guy could unwind and talk about the day's activities.

Arm Wrestling at Johnny's Corner

Arm wrestling was one of the favorite activities at Johnny's Corner. Construction workers are typically pretty macho, and some of them are pretty darn big and in good shape. So it was not unusual to walk in and see a match underway with guys standing by the side watching and making bets on the outcome.

My office and materials testing laboratory was only a mile or so from Johnny's Corner. One day Bobby, one of my technicians, glanced out my office window at the ready mix concrete plant across the street. A black worker was lifting a 55-gallon drum of liquid additive and putting it on the pallet to be picked up by a forklift. He clamped his hands on the side of the drum, and raised it onto the pallet using only his arms. He didn't even bend from the waist. The way he lifted it, it looked empty.

Bobby went over to the plant. For the hell of it checked the level of liquid in that 55-gallon drum. The drum was ¾ full, which was about 35 gallons, which was about 280 pounds. The guy had lifted almost three hundred pounds with no effort, using only his arms.

The worker was named Levi. He was an African-American who was born and raised in Florida. He was not especially big and did not look real powerful. He was solid and lean, but his arms did not look big like those of guys who pump iron. He was a good family man who went home each night to his wife and kids. Levi had never been in Johnny's Corner. In fact, he told us he had never touched alcohol in his life.

We convinced Levi that he could make some serious money arm wrestling. We showed him how to do it to make it look like he was winning but only with great effort. He came over to the lab after work every night for a few days and practiced with Bobby or me until he could make it look good. To this day, I have never met a stronger man than Levi. He also proved to be of Sidney Poitier quality as an actor.

Friday night was the chance for Levi to show his stuff. The first guy challenging all comers was Tommy, a burly ironworker who was pretty good at it. Tommy put down a couple of guys with no trouble, and then it was Levi's turn. As far as I know, Levi was the first black guy to arm wrestle in Johnny's Corner.

Levi and Tommy locked hands and started to grunt and strain. Levi's arm wobbled then went down close to the surface of the table, then recovered. Then Tommy's arm went down, back and forth. Finally, with what looked like a lot of effort, Levi put Tommy down. We collected our money. Another challenger came along, and Levi put on the same show and won again.

By this time Bobby and I had turned $100 into $1000 and were waiting for the inevitable. In walked Billy Ray Conroy. Billy Ray was a laborer, 6 foot 6 inches tall and about 300 pounds. He looked like an NFL defensive lineman, but with a dumber look in his eyes. He had a low forehead and scraggly brown hair in a cheap bad haircut. His massive forearms were hairy and tattooed. His right arm had a tattoo of a cobra coiled around it and ready to strike. His left arm was decorated with a Florida State Seminole football helmet and a couple of dozen tomahawks. His biceps were the size of a normal man's thighs. He was mean and smelled bad and had never been beaten at arm wrestling.

Levi had taken down about five guys when Billy Ray came smirking up for his turn. Bobby and I spread our $ 1,000 over the room and had no trouble getting plenty of guys to take Billy Ray. Levi hadn't looked that impressive taking down the clowns earlier in the evening, and he was a good 8 inches shorter than Billy Ray and at least 100 pounds lighter.

The two guys faced off, locked hands, and struggled for almost ten minutes. They grunted and wobbled and dripped sweat all over the table. First Billy Ray's big arm would push Levi's arm halfway down, and then Levi would do it to Billy Ray. It looked so convincing that Bobby and I weren't sure if Levi could win. Then slowly, grunting and snorting with apparent great effort, Charlie finally put Billy Ray's arm down.

Billy Ray's face turned white, the room turned quiet, then all hell broke loose. The whole room was screaming and shouting in surprise. It was Billy Ray's first defeat. We collected on our bets, and now had $ 2,000 to be split 50-50 between Levi on the one side, and Bobby and me on the other. We weren't giving or taking odds, just a straight bet, win or lose.

Billy Ray panted and gasped and rubbed his shoulder and glared at Levi with his dull little eyes.

"It was a goddam trick of some kind," he said "Goddamn you, you took me by surprise. I want to do it again, you black sonuvabitch!"
Levi looked reluctant, but agreed to arm wrestle Billy Ray again. They faced off again, and this time we could only find $ 500 in the room ready to bet against Levi. This time when they locked hands, Billy Ray immediately put all of his power and massive body into the attack. Levi held firm with no emotion on his face. After about 5 minutes, which seemed like an hour, with Billy Ray straining and sweating and cursing, Levi yawned and looked

at the beat up old Mickey Mouse watch strapped to his left wrist.

"I gots to be getting on home," said Levi.

Then, with no more effort than a small girl setting a book down, he gently forced Billy Ray's hand down on the table.

The spectators yelled and screamed and cheered. Billy Ray sat sobbing softly at the table, rubbing his sore shoulder looking like a school bully who had just been whipped by the smallest kid in class. Bobby, Levi, and I went out into the parking lot and split the money. I don't know if Levi ever told his wife how or where he made that money, but I'm sure it came in handy.

Same Area in 2014. IHOP Is Where Johnny's Corner Was Located. Google Earth Aerial Photo

Johnny's Corner is long gone, replaced by a Texaco convenience store and gas station. Jammed around the intersection where scrubby groves once slept in the sun are an Olive Garden, Macaroni Grill, a Sizzler, Chili's, Shoney's, and plenty of places to sleep like Country Inn and Suites, Sheraton, and Courtyard. The massive bulk of the luxurious Grand Cypress Hotel looms in the sky just to the west, and a short hike south the hotels of Lake Buena Vista reach high into the sky.

The only reminder of Vineland still standing are a few old homesteads and a school house that was built in 1950. The building is an old single story brick building that was there back before Walt Disney World was built.

Shortage of Housing for Workers At Walt Disney World

One of the first problems I faced was where to put up the guys until they could find permanent housing. The area around Disney World was largely deserted, and what few little motels existed in the area were booked solid. I found several rooms at Horne's Motor Lodge in Orlando. Horne's was 17 miles from the Magic Kingdom. There were almost no other motels between Walt Disney World and Horne's along I-4. It wasn't until 1970 that the Hilton and others began to sprout up along the brand new International Drive which paralleled the interstate between Kirkman and Sand Lake Roads.

Most of my technicians stayed at Horne's, at least for a while until they could find permanent housing. All kinds of housing developments and apartment projects were under construction in anticipation of the opening of Walt Disney World. But until they were finished, there was a real housing shortage in Orlando.

The Search For Qualified Construction Material Testing Technicians

Bill Moss and I visited some of the steel fabrication shops that were producing items for the Magic Kingdom. Bill wanted to demonstrate to these shops that Disney was paying for quality products, and would accept nothing less. We quickly replaced all of the inspectors of the northern firm that Disney had fired with Law Engineering Testing Company staff. Soon the shops were producing quality steel again. But we had been borrowing technicians from Law's offices all around the southeastern United States, and needed to hire some new people so these guys could go home.

It was tough back in 1969 and 1970 to find qualified inspectors and testing personnel to work on the project. The aerospace industry was taking a dive since landing the man on the moon so thousands of aerospace engineers and technicians were on the market. Unfortunately, almost none of them had the kind of hands on experience in construction needed for our work.

Most of these aerospace workers were extremely specialized, with titles that did not really describe what it was they did: digital analysts, program specialists, fuel technicians, and payload specialists. Many of them had titles which said they were engineers, but they had no college education in engineering.

My theory has always been that our space program reflects the Germans who set it up. Dr. Werner Von Braun, Kurt Debus, and others were liberated from Hitler's Germany after World War Two and brought over here to get us into the space age. These Germans were so couched in secrecy that they fragmented the tasks so very few people would ever know what the big picture was.

I'm sure there were some very competent people involved in our space program, or we never would have made it to the moon. George Knudsen, also known as Mechanical George, was a good example of an engineer who helped put the man on the moon and also helped build Walt Disney World. Others from the space program did not fare so well.

CHAPTER 7. WALT INSISTS ON HIGH QUALITY CONSTRUCTION

This aerial view from Google shows the Magic Kingdom in 2017. The area of Vineland and Johnny's Corner is out of sight just past the lower right hand corner of the photo. By the end of 1968 the dredging of Bay Lake was well underway and much of the dirt in the area of the Lagoon had been moved into various sections of the Magic Kingdom Park.

The Magic Kingdom, 2017, Google Earth Aerial Photo

Bay Lake was a natural body of water; the Seven Seas Lagoon (to the left of Bay Lake in the photo) was excavated to form a water body and the soil was transported to the Magic Kingdom Park as fill. Much of the soil was also used to bury the structures under the Magic Kingdom that have become famous as the tunnels or utilidors.

Disney Hires Construction Quality Control Firms

The Disney organization was very concerned that Walt Disney World be of high quality construction and absolutely safe for the guest. By 1969 they had at least three firms involved in various aspects of the construction quality control program.

Dames and Moore was in charge of soil testing and evaluation of all foundation materials and designs. They had been heavily involved in mapping the soils on the site so that the proper materials could be used where needed. They monitored all of the earth moving and filling activities.

Law Engineering Testing Company handled testing and inspection of all concrete and asphalt on the project. Disney had a ready mix concrete plant operated by Rinker Materials on the site; Law had a full time inspector at the concrete plant and technicians that sampled and tested all concrete from ready mix trucks as they poured concrete in the various structures.

A northern firm was in charge of steel construction quality control, including welding, bolting, and inspection of fabrication shops and field erection. There was not enough steel fabricating capacity in Florida to handle the huge steel demand of Walt Disney World. Shops from all over the southeastern United States and Pennsylvania complemented the shops in Florida.

Haunted Mansion and Small World were under construction in Fantasyland, and Disney was extremely upset over the poor quality of some of the fabricated steel that was being shipped to the project. The northern firm had a full-time inspector in the Atlanta shop that built the girders, so they were blamed.

Disney fired them and asked Law Engineering Testing Company if they could also provide staff to handle the steel quality control. I worked for Law at the time in their Tampa branch and had previously worked for a time at Bethlehem Steel and was a certified welder and graduate civil engineer.

Disney accepted my qualifications as manager for steel quality control. Law began to comb its branches around the country to transfer technicians to the project that were skilled in steel inspection, welding inspection, non-destructive testing, and general steel quality control techniques. Dozens of inspectors were transferred to Disney by Law and dozens of others were hired throughout the United States and brought to the project. Dozens more worked in steel fabrication shops in Florida, Georgia, Alabama, South Carolina, North Carolina and Pennsylvania.

I soon became certified not only as a welder, but as a radiographer and ultrasonic technician. I became **Radiation Safety Officer** responsible for the safe use of the isotopes that were used in radiographing welds. I also managed the inspection and testing staff on the site and in the various fabrication shops around the country.

Finding housing was a problem for the thousands of construction workers and technicians who labored on the project. Many of them were from other parts of Florida or other states. Some stayed in local campgrounds or parks. Other stayed in one of the few Orlando motels close to the site during the duration of construction. A typical motel of the time was Horne's Motor Lodge at the intersection of Interstate 4 and South Orange Blossom Trail (called SOB Trail by the locals).

Horne's was a new place with decent rooms and a restaurant on site. The motel was part of a chain that had quite a few outlets in the South. It was the closest major motel to Disney World 17 miles to the southwest.

CHAPTER 8. EXAMPLES OF DISNEY QUALITY

Walt and Roy had a sincere concern for the safety of their employees and guests. This concern resulted in an extremely high commitment to quality in construction. Walt Disney World may have been the highest quality construction project in the history of the world

Friends of the Disney brothers remembered that back in 1938 Walt and Roy bought a house in California for their parents. A few months after Elias and Flora moved in, a bad furnace filled the house with carbon monoxide, and Flora died as a result. Walt and Roy blamed themselves because they had bought the house. This tragedy may have been the beginning of their life-long focus on quality.

Here are some examples of the Disney demand for quality.

The Tampa Shipyard Welding Incident

A shipyard in Tampa was building the large passenger ferries that would be used to carry guests from the main entrance to the Magic Kingdom. My technicians at the shipyard discovered that most of the welds on the ships were not full-penetration as designed and required by Disney specifications. At first the yard challenged our findings and said we were too picky. When that didn't work, they got a naval architect, the one who had designed the ships, to say that full-penetration welds were not needed. They also said our unreasonable quality standards would make them late in delivering the ships.

I traveled with Admiral Fowler and other Disney executives to a meeting at the shipyard. The purpose of the meeting was to show our radiographs of the welds to Admiral Fowler. The shipyard's naval architect would then make his pitch to the Admiral for relaxing the standards.

When the presentation was over, the room waited for Admiral Fowler to say something. I was scared to death he was going to say that my boys were too picky, that we were nitpicking the project to a late completion. The Admiral had been quiet during the entire event, but finally he stood up and looked at the president of the shipyard.

"This reminds me of my Navy days," the Admiral said. "It was back in the 1930s. We were building the first welded submarine hull. Up until that time, the hulls had been riveted."

He paused and walked over to the illuminated viewer where one of the radiographs was still displayed. He stared at the radiograph for a long minute.

"Welding technology was new back then, very primitive, very much trial and error. We had never done it before, so we had to learn as we went along," he said.

Then he turned around and looked at the president of the shipyard.

"What we were doing back then in those early days was a lot better welding than what I've seen on these radiographs here in Tampa in 1970. I want you to tear out these welds, do them again the right way, and do whatever you need to do to get back on schedule at no expense to Disney."

Admiral Fowler turned sharply and left the room. We all followed him like ducklings walking behind their mother. We were all glad he was on our side. We knew he had done what Walt would have done.

The Monorail Beams

One example of the Disney commitment to quality and safety can be seen in the incident of the monorail crossheads. The monorail beams were pre-stressed concrete sections that were built in the Pacific Northwest and shipped to Walt Disney World by rail. The beams were extremely long and three flat bed railroad cars were needed to handle each beam. This arrangement could only take gentle curves, so the rail route across the country to Orlando had to be carefully planned to avoid any sharp curves. Routing was especially tricky through the mountains.

The beams arrived in Taft, a small town south of Orlando where the Strates Carnival trucks and train cars spent the winters. They were then hauled to Walt Disney World along Taft-Vineland Road on big trucks that had steering tractors on the rear. Each truck had a driver up front and a tractor driver in the back.

Once at the site, the beams were supported by tall vertical concrete columns, which circled the area from the Main Entrance and ticketing area to the Magic Kingdom. Each column was topped by a thick steel vertical plate designed to handle the transfer of stresses from beam to beam. This steel member was called a crosshead plate, and it was custom made of very thick high strength steel rolled from an ingot in the steel mill.

My ultrasonic technicians inspected each plate when it was delivered to the project from the steel mill. Of the several hundred we inspected, a few had de-laminated zones. These were flat, squashed areas within the plate that could possibly be a weak spot. This condition sometimes happens when the molten steel ingot has a gas bubble in it before it cools. When the ingot is rolled out into a thinner section, the bubble becomes flattened and elongated, and remains within the plate after the steel cools.

The crosshead plates were on the critical path of construction. If the steel mill had to make new ones, the project could be delayed. The opening date of October 1971 was getting closer and closer. Disney was not about to take a chance by using defective materials, so something had to be done.

Bill Moss, John Zovich, and Don Edgren met with me one day and instructed me to find the "world's leading expert on delamination in steel plates loaded under torsional stress". They wanted an expert whose opinion would be indisputable, and they didn't care how much it would cost to find such a consultant. They wanted an objective opinion, even if it meant bad news, but they wanted an opinion that everyone could believe in. And they wanted it fast.

I searched technical journals and called committees of the American Society of Mechanical Engineers and the American Society for Testing Materials. I called professors at leading universities and researchers in various steel institutes and organizations. My search led to a Ph.D. employed by a private firm in Massachusetts. Every one of the experts I talked to recommended this Ph.D.

Disney retained the good doctor, and he came down to the project immediately. He examined all of our ultrasonic plots, and asked for some additional tests so he could witness them. He concluded that all of the delaminated crosshead plates could be used by putting them in areas where the stresses would not affect the delaminations. Disney accepted his opinion. Months of delay were avoided. The plates have performed safely since Walt Disney World opened in 1971.

Holiday Detection By Jeeping All Piping Systems

All of the Disney piping systems that were installed underground were painted with a protective coating before back filling with soil to protect them against corrosion. We checked these coatings with devices called Jeepers. They looked a bit like a long handled metal detector, the kind for finding treasure. At the business end they had a heavy coil spring which you unhooked and snapped around the pipe barrel. Then you pushed it along the pipe. If there was a hole or gap in the coating (a holiday), it would set up a spark indication and we would mark it for later covering. This was called officially, holiday detection.

Disney's Acronym Companies

Walt and Roy created many companies over the years to handle special projects. Many of these names were known by their acronyms. **MAPO** was short for Mary Poppins. **WED** stood for Walter Elias Disney. **RETLAW** was Walter spelled backwards. **BVCC** was Buena Vista Construction Company. **WDW** was Walt Disney World. **RCID** was Reedy Creek Improvement District. **EPCOT** was Experimental Prototype Community of Tomorrow.

Only the Best Welders Could Work On Walt Disney World

We also certified all welders. Disney insisted that all welders be specially certified for their project.

We had the welders prepare samples under our inspection, then tested them. This is where I learned that it is possible for someone with 30 years experience to really have 1 year of experience 30 years in a row. Number of years of experience was not always an indicator of who was a good welder and who wasn't.

Disney's Major Welding Radiography Program

Each joint of all butt-welded piping systems was radiographed. We used a radioactive isotope, **Iridium 192**, to expose the film. The isotope was in a small pellet, the size of a peanut. The pellet was at the end of a long cable, like a plumber's snake, and the cable was within a lead-shielded tube. When not in use, the pellet was retracted into a heavy lead container the size of a bowling ball. I was the **Radiation Safety Officer** for the project, having gone to a special school and obtaining the appropriate certification.

To use the isotope, the technician stood a distance from the lead container with a cable handle in his hand. He then cranked out the isotope so it popped out of the shielded tube at a point immediately over the welded pipe joint. The pipe joint had been covered with radiographic film on the side opposite of where the isotope was. The film was then exposed by radiation for a few minutes, the pellet pulled in, and the film retrieved. Every once in a while the pellet would hang up or come loose out at the pipe. Then we had to retrieve it using a method of **"splitting doses"**.

Each of us who worked in radiography wore a film badge that recorded cumulative doses of radiation. We turned in the badge weekly and our doses were recorded. If you received too much, you couldn't work. We also each carried a dosimeter. This was a device the size of a pen which measured radiation on a daily basis. If you got too much in a day, you might have to sit out for a while.

Splitting doses was a way of sharing the risk of retrieving the hung up isotope. The first man would run out to the pellet with a pair of long-handled tongs and flip it back towards the lead container. The second man would then run out and flip it the rest of the way to a point immediately adjacent to the container. The third man would run out and stick the pellet in the container. Each man would receive more than a daily dose, sometimes a weekly dose, but no one man would be out for too long.

Unions Try to Replace Steel Inspectors with Their Members

My steel inspectors all carried torque wrenches that were used to make sure that the high strength steel bolts on structural connections had been properly tightened. The wrenches had a dial and were calibrated on a machine known as a Skidmore-Wilhelm Bolt Tension Indicator. This device correlated foot-pounds of torque read on the wrench dial with actual tension in the bolt. We torqued (that is tested) all high strength steel bolts on every structure at Walt Disney World.

Early in the project the business agent for the ironworkers union passed word to me that my technicians were violating union rules by torquing high strength high tension bolts. He said that only ironworkers could do that.

I passed the complaint on to Disney, and they convinced the agent that it was not a violation for my trained technicians to do the work.

Admiral Joe Fowler and His Fleet of Ships

A large fleet of ships and boats would be needed to transport guests from the ticketing area across the lagoon to the Main Street entrance. A shipyard in Tampa built a couple of the larger vessels out of steel.   My technicians were stationed at the shipyard and radiographed all of the hull welds on these ships.

Morgan Yacht Corporation in Clearwater made some of the smaller boats of fiberglass.   My technicians were also there, inspecting their construction techniques to ensure that the boats were being built to Disney standards.

A service area for maintaining Disney's fleet of ships was built on the northeast corner of Bay Lake. The facility included a dry dock, where the larger vessels were brought periodically for servicing, maintenance, and repair. This service area was a favorite of Admiral Fowler since it reminded him of his days in the Navy. He had talked Walt into building one at Disneyland, and had taken a lot of teasing from Walt about what Walt called **"Joe's dry dock"**.

One day down by this ship service area the Admiral was personally involved in straightening out some problem. Things weren't going his way. He got madder and madder and finally stormed out of the site, jumped in a pick-up truck, and drove back to his office. He did not realize he was in the wrong truck. All the trucks looked pretty much alike, white Fords with the Disney logo. We all left our keys in them when we parked in case they had to be moved. The poor guy whose truck Joe took walked back to the office. He didn't have the nerve to drive Joe's truck.

Cement Poisoning Is Not A Pretty Thing

My technicians worked in the field, but we also had a testing laboratory on site. This is where we cured and tested concrete cylinders, examined radiographs, and did welder certification and soil testing.

One day I heard screaming from the lab and one of my guys rushed in saying something was wrong with Charles, a concrete technician. I ran on out to find him writhing on the floor in obvious pain. His head was swollen to the size of a basketball, and his hands looked like catcher's mitts. He was a skinny kid, but this day his face was bloated like a balloon with a painted on face. The ambulance came and took him to the Disney clinic located in Lake Buena Vista.

The doctors said it was cement poisoning. When they work with cement for a long period of time, some folks get an allergic reaction. Charles was one of those unfortunate people. We reassigned him to other work and kept him away from cement and concrete.

The Swiss Family Tree and Dr Phillips Packing House

The **Swiss Family Tree House** was made of steel plate and pipe and built on site at the Magic Kingdom. This structure was then wrapped in fiberglass under the supervision of the Disney art directors, and it looked very much like the bark on a Banyan tree.

An old ramshackle wooden abandoned packinghouse sat near where Sand Lake Road now hits Apopka-Vineland Road, several miles north of the Magic Kingdom. This old building along the railroad tracks sheltered unskilled workers who sat day after day under the faded tin roof of the building tying the leaves onto narrow steel rods.

The rods and leaves would later be attached to the tree at Walt Disney World.

Like everything Disney did back then, the leaves were tied meticulously and looked very realistic.

Swiss Family Tree House

Walt Disney World's Famous Tunnels Are Not Tunnels

During the first quarter of 1970, mountains of dirt had been moved out of the lagoon area onto the Magic Kingdom site. Before the earth moving began, Fantasyland Basement was built at ground level on the site.

The entire Theme Park was being raised about 16 feet so that the basements would be buried beneath the soil and out of sight. The roof of the basement would become the area that the thousands of guests would walk over while going to the various attractions. The dirt came from the lagoon.

The Wave Making Machine Fails At The Polynesian Resort

Not all of Disney's projects at the Magic Kingdom were successes. One that stands out in my mind is the wave-making machine. This large device was built on an island in the Seven Seas Lagoon offshore of the Polynesian Resort. It was a huge device with blades and paddles and hydraulic rams. The machine was designed to push large waves of water onto the beach at the Polynesian to allow for surfing and a more natural island ambience.

The Polynesian Hotel featured a luau every night, and Disney hoped the waves would add to the guest experience. The machine got cranked up several times, sent a few feeble waves toward the beach, then usually broke down. It would have been comical had it not been so serious.

One story that made the rounds told of a prominent Disney boss, dressed in surfing shorts and paddling around in the lagoon on a surf board waiting for the big wave that would carry him triumphantly up to the beach. The wave never came, and the Disney folks finally gave up on the wave-making machine and went on to other ideas.

CHAPTER 9. THE 7TH PRELIMINARY MASTER PLAN

Although Walt Disney died in 1966, his concept of Walt Disney World was complete.

He came to a meeting late in 1966, only weeks before his death, with a plot plan for the Magic Kingdom which was very close to what was finally built. The plan was in his own hand and drawn on a breakfast napkin.

Contemporary Hotel and Main Parking Lot 1971 - Lagoon In Left Center, Bay Lake Upper Center
State Archives of Florida

This document was known for the rest of the project as the **7th Preliminary Master Plot Plan.**

The plan showed the Magic Kingdom in conceptual detail, and was followed after Walt's death with very little change.

The plan showed **The Magic Kingdom** on the far northern end of the 27,000-acre tract. Just south of it would be a new lake named **Seven Seas Lagoon**, and just east of the new lake would be a restored and improved **Bay Lake**. The **Contemporary Hotel** would be located between Bay Lake and the Lagoon.

The Contemporary Hotel Under Construction. Circa 1971 - Bay Lake In Foreground, Lagoon At Top
State Archives of Florida

Bay Lake was connected to the Seven Seas Lagoon by a viaduct that passed over the road leading to the Contemporary Resort. It never fails to fascinate me when I see large vessels passing overhead as I drive under that viaduct.

Guests would enter the Magic Kingdom from the south through Main Street USA. The guests would arrive at Main Street from the ticketing area either by boat across the Seven Seas Lagoon or by monorail. Once they had arrived at Main Street, they could either walk into the Magic Kingdom through Main Street or take an open train that encircled the kingdom and made stops along the way.

The Magic Kingdom consisted of Main Street USA; Adventureland; Frontierland; Liberty Square; Fantasyland; and Tomorrowland. The main view down Main Street was of Cinderella's Castle, which dominated Fantasyland. Each land had its own unique restaurants and rides.

Construction Begins on Lagoon, Bay Lake, and Magic Kingdom

Bay Lake was a typical Florida swampy water body with a bottom of mucky, dark, organic root material. The lake water was the color of tea. Disney dredged out the mucky stuff, and found beautiful white sand beneath it. They took the sand and created wonderful beaches all around the shores of Bay Lake.

Just west of Bay Lake was a low area that was excavated and made into a new lake called **Seven Seas Lagoon**. The sandy soils that came out of the lagoon were used to backfill around the structures in the Magic Kingdom.

The "basement" level of all of the key areas in the Magic Kingdom was built first. Heavy reinforced concrete construction was used, and after the basements were waterproofed the buildings in the Magic Kingdom were built on top of them. The lagoon soils were then placed around and on the basement structures, creating the famous Disney tunnels or utilidors.

This subterranean empire allowed Disney workers and service vehicles to stay out of site of the guests so the illusion of fantasy would not be disturbed. Utility lines were exposed in racks suspended from the walls and ceiling of the utilidors. Since they were accessible this way, it meant that streets would never have to be dug up to fix things.

All of the systems servicing the various attractions within the Magic Kingdom were in these utilidors. High temperature hot water, gas, water, wastewater, and a vacuum operated solid waste collection system called **AVAC**. My technicians inspected the welding of all of these systems, and radiographed all pipe joints. They also inspected the installation of the systems in the racks, and the installation of the racks.

I signed many thousands of reports certifying to the integrity of these systems. To the best of my knowledge, all was okay then and still is now. But what if one of my technicians had a momentary lapse of judgment and missed a huge flaw in the high temperature hot water line? What if that line burst on a shift change when hundreds of employees were down there? Worries like that make a young engineer's hair turn white before his time.

Cinderella's Castle Under Construction Circa 1971

Cinderella's Castle sat on top of what we referred to as **Fantasyland Basement**, but which is one of the typical Disney underground areas. The castle frame is of structural steel, mainly bolted connections, and is sheathed in fiberglass. The employee cafeteria and changing rooms are down in the Fantasyland Basement.

One day shortly before grand opening I sat next to **Snow White** while she ate lunch. Another day I chatted with **Mickey Mouse** as he shoveled in his lunch. Mickey's head was sitting on the table next to his plate. I noticed that the head had a little battery fan and dry ice in it to keep the inside cool during the hot Florida days.

Up near the top of the Castle, behind a parapet with gothic towers, is **Walt Disney's apartment**. It was originally going to be built for Walt. After Walt died, the company planned to finish it for his widow. When she decided she didn't want the apartment, it was not completed. It was laid out, and rough plumbing and electrical wiring and outlets were installed. The apartment was never finished, at least not during my years on the project.

The Myth of Walt Disney's Frozen Body

The apartment floor had a spectacular view of the entire Magic Kingdom, Bay Lake, Lake Buena Vista, and the thousands of acres of citrus surrounding the entire property. On a clear day you could see downtown Orlando. I still think it would be great to live in an apartment in Cinderella's Castle. I would love to have that apartment finished and live in it to this day and I imagine Walt was looking forward to it also.

I believe this apartment is where the story started about Walt's body being frozen. I remember hearing someone say that a massive refrigeration system and cryogenic chamber were installed in Cinderella's Castle. The rumor was that Walt was placed in a glass box and kept frozen like a fish in the freezer in your home. When a cure for cancer was discovered, he would be thawed out and cured. I wouldn't be surprised if some of the more imaginative construction workers – or **Imagineers** - started the rumor.

The Administration Building and Pluto Park

The first permanent building I can remember being built at Walt Disney World was the **Administration Building**. It was just a little south of Winter Garden -Vineland Road. This building housed the WED staff and some of their consultants. Admiral Fowler and his staff had office there. The building also had a cafeteria, because there were very few places to eat near the project site back then.

The building was a pre-engineered model, manufactured by Butler, shipped to the site and erected by contractors supervised by Disney. It was a pretty big building, about 100 feet wide and three times as long.

The next major complex that was put up at WDW was what we all called **Pluto Park**. This was a huge village of individual construction trailers arranged and connected by wooden walkways that served as the headquarters for field personnel for Disney managers, consultants, and contractors.

Quality was always first in the Disney attitude. Structural steel was an example. The steel had to meet strict standards established by the American Society for Testing Materials. We had to be able to match mill certificates showing the metallurgy of all steel to the actual steel that was delivered to the site.

The unions had succeeded in forcing Disney to use only steel made in America. The requirement was written into the project specifications. We began to have trouble finding enough American steel that met the standards, so the unions finally relented and we began to use steel made in Spain and Belgium and Japan.

A large ready mix concrete company set up a plant on the Disney property to handle the huge volume of material needed to build the project. My technicians tested all of the concrete that was placed for compressive strength and other qualities.

Disney Buys a Railroad Bridge

Innovation and creativity were the hallmarks of the Disney Imagineers. They knew of Walt's love of railroads; he had a miniature railroad at his home in California and loved to play engineer and give neighborhood kids train rides. The Imagineers decided to add a special touch to the Main Street Railroad that encircled the Magic Kingdom. They would add a real drawbridge. They ended up buying a used one from the Florida Department of Transportation (FDOT).

Wabasso is a small village on the western shore of the Indian River 7 or 8 miles north of Vero Beach. An ancient iron drawbridge crossed the Indian River from Wabasso over to Orchid Island, the barrier island that forms the beach area north of Vero Beach. Such upscale communities as Johns Island, Sea Oaks, and Windsor are now on this island. Back in the early 1970s it was largely unpopulated.

Florida DOT was replacing the old drawbridge with a modern high rise concrete bridge. Disney Imagineers looked the old bridge over and arranged to buy it from FDOT. The Disney engineers supervised cutting the bridge into smaller sections, and these were then transported by barge to Tampa by way of the Okeechobee Waterway which runs across the state from Stuart to Fort Myers. A shipyard in Tampa put the bridge members back into good and safe condition.

The bridge was then transported by truck to Walt Disney World, where it was assembled and is now part of the **Main Street Railroad**. The bridge is functional, and is opened periodically to allow large vessels from the lagoon to enter into the dry dock area on Bay Lake for repairs.

CHAPTER 10. SOME PEOPLE WHO WORKED ON THE PROJECT

Much of the success of a big construction project depends on the personalities and dedication of the players involved. About 9000 people were involved in the design and construction of Walt Disney World, and I only worked with a handful of them. It is impossible to list all of them, or even to remember who all of them were after all of these years, even the ones who had key roles and deserve to be remembered.

Florida had a much smaller population when construction began on Walt Disney World than it has now. There were not enough skilled construction workers in the Orlando area, maybe even in the entire state, to pull off a project the size of this one. Roy Disney had told his troops they could spend no more than $ 125 million. By the time they got it finished the way they wanted it, the cost was probably more like $ 400 million. And that was in the late sixties and early seventies, before the expense of LBJ's Great Society inflated the cost of everything by a factor of up to ten.

To get the people they needed to build Walt Disney World, the union was brought in. Union guys from all over the country came to the swamps and piney woods of central Florida to help build the project. These guys were called **"boomers"**; they always followed a construction boom, wherever it was happening. A lot of them got sand in their shoes, as they say in Florida, and stayed in the state.

There were too many of them to single out; ironworkers, carpenters, operating engineers, millwrights, electricians, mason, and laborers. Each of them had their part to play.

The construction of Disneyland in Anaheim had suffered from union created work stoppages. For Walt Disney World, a different contract was used which promised the union more jobs and a quick dispute resolution method. In return, Disney asked the unions to cooperate and prevent work stoppages. The main players, especially the ones who worked for the Disney organizations, are the ones I saw a lot of and who had a lot to do with my life on the project. I will never forget these people. They played a big part in my post-graduate real-world education, and helped give me some of the knocks that I received in the college of hard knocks.

Many of the Disney people I worked with had worked directly for Walt Disney during the construction and operation of Disneyland. When these people were faced with a major design or construction problem, one of them would always ask: **"What would Walt do?"** Usually, they knew what Walt would have done and proceeded according to what they believed Walt would have wanted.

Walt Disney had insisted that everyone in his organization call him Walt, not Mr. Disney. That tradition carried down and was alive and well when I worked on the project. It was always hard for me, still in my twenties at the time, to call these older, experienced, awe-inspiring men by their first names.

Walt's Big Brother, Roy Disney

Roy Oliver Disney was Walt's older brother. Walt always looked up to Roy, and they became a team shortly after Walt started the business. Walt and Roy were a perfect team. Walt had the dreams and could not care less about how much it cost to achieve his dreams. Roy was the financial wizard who figured out the ways to make Walt's dreams into reality. He had a way of convincing bankers to back the Disney projects. Roy didn't always agree with Walt, but Walt usually got his own way.

After Walt died in 1966, Roy picked up the total Disney dream and charged ahead. Roy insisted on renaming Disney World. From that point on, it was known as Walt Disney World. Roy wanted the world to remember whom it was that created the place. He was a small, friendly man, bald with glasses, who looked like you'd want your favorite uncle to look like. He was around the project continually during the years I worked there. He seemed to have an active part in what was going on at all times. He usually wore a business suit, and was easy to spot when he was out on the construction site.

One day I was in one of the utilidors inspecting some work when Roy Disney walked up to me. He smiled at me and said, looking at the racks of pipes and electrical conduits neatly arranged overhead,

"This is what the inside of a space ship must look like.", he said.

Admiral Joe Fowler

Joe Fowler was in charge of building Walt Disney World, and also had a lot to do with designing it. He was a retired U. S. Navy Admiral, and had built many shipyards during World War Two. Walt hired Joe Fowler to be in charge of constructing Disneyland in Anaheim. Fowler got that done and even ran the place for several years after it opened. He had the nickname of **Admiral "Can Do"**, because every time Walt asked Joe if he could do something his response was always "Can do, Walt".

The Admiral was a natural, forceful leader who was very much in charge of building Walt Disney World. He was always clear and forthright in telling his staff what he wanted. Fowler was in his seventies in 1969, a tall powerful man with a bald head and a white fringe of hair. He was usually easy going, but could be very caustic and volatile when pissed off. He knew how to inspire people to get a project done on time and in the right way. He had helped whipped the Japanese and the Germans, and could do the same damn thing to any obstacle thrown in his way.

A California construction company was doing most of the work at Walt Disney World. Apparently they told Joe Fowler sometime in 1970 that they would not be able to finish by the designated drop-dead opening date of October 1, 1971. The following

Friday, Joe fired the contractor and had Disney buy all their equipment. The Disney staff scrambled over the weekend to make the transition happen.

On Monday, a newly created Disney construction company was in charge of the site work. **Buena Vista Construction Company** burst on the scene led by Disney managers and proceeded to finish the project.

Walt Disney in Cap, Roy Disney, Disney President Card Walker, Joe Fowler
State Archives of Florida, Circa 1966

General Joe Potter

William E. "Joe" Potter was also involved on a daily basis. Joe was a retired U. S. Army General. He had spent much of his career with the U. S. Army Corps of Engineers, and understood large projects. He had been governor of the Panama Canal Zone at one time in his career. He had worked for **Robert Moses**, the New York City master builder, on the 1964 New York World's Fair. Disney built pavilions there, and Walt was impressed with Joe Potter's ability. When the project was done Walt hired General Potter to come back with him to California and work for him on **Project X**, which later became Walt Disney World.

General Potter was a short, stocky man with wavy grey hair. He had a steely look behind his smile that left no doubt that he was a commander. With or without a uniform, it was apparent that he was a guy accustomed to being in charge.

After the Florida land acquisition and Walt's death, General Potter moved to Florida and was instrumental in the planning of **Reedy Creek Improvement District** and getting it approved by the Florida legislature. He became highly visible to the people of Orlando long before Admiral Fowler did. Before construction began on the Magic Kingdom, General Potter organized and supervised the drainage project and getting the site ready for construction.

After Joe Fowler moved to the site, it was General Potter who intercepted the media, the politicians, everybody who could distract the Admiral from his duties in getting the project built. As far as the public was concerned, he was much better known than Admiral Fowler was. It looked to the outside world like the General was in charge of everything. He and the Admiral worked like hand in glove, and pulled it off very well.

These three men were the biggest and brightest stars in my firmament, but they were surrounded with some of the most competent, hardworking, dedicated and imaginative engineers and construction managers that have ever been assembled.

Walt Disney and Imagineer Don Edgren

Don Edgren

Don Edgren was with WED Engineering, and was the top WED guy full time on the site. He moved from California to Florida to handle the engineering on the project. Like many Disney executives of the time, he lived in the little town of Windermere, a few miles from the project. Don was a laid back guy, but very practical and very persistent in getting what he wanted. He reminded me of a technically inclined Bing Crosby.

John Zovich

John Zovich was another WED engineer, a hard charger, and a guy who could get things done. He had a business like flat top hair cut and was a heavy smoker. He shuttled back and forth between the WED design offices in California and the project in Florida.

Bill Moss

Bill Moss was my immediate client contact during the project. Bill was a civil engineer with bachelor and masters degrees in civil engineering from the University of Florida. One of his jobs was to supervise the activities of all the testing laboratories and inspectors on the project. That included my staff and me.

I had known Bill earlier in my career in Tampa. Bill was a structural engineer with Watson and Company, and worked on the design of Tampa Stadium in the late 1960s. I was with Law Engineering and we did the test borings and construction materials testing on the stadium. Bill was a wiry guy with a demanding nature who expected top service from his consultants and got it or got rid of the consultants.

Pat "Patty" Branham was Admiral Joe Fowler's executive secretary and attended most meetings the Admiral attended and took notes and produced minutes.

Ossie Tanner

Ossie Tanner was a flat topped technician who was an extra pair of eyes and ears for Bill Moss out on the site. He was very knowledgeable about construction techniques, including soils, concrete, asphalt, and structural steel. Ossie stayed with Disney as a construction consultant for years after the theme park opened working on new projects like Epcot and Animal Kingdom.

Arnold Lindberg

Arnold Lindberg was a powerful flat topped extrovert in charge of Disney's machine shops and ground transportation system. He went back to the old Disneyland days and had lots of great stories about Walt Disney and Joe Fowler.

Arnold liked to blow off steam after work at a place north of Orlando, in Fern Park, called Freddie's Steak House. It was at that time maybe the best restaurant for steaks in Florida. Freddie's had a piano bar, and a pretty lady who played the piano. It was at Freddie's that Arnold earned the title **"Swedish Nightingale"**. The gal would play and everybody around the piano would sing. Arnold's strong baritone and friendly forceful extroverted personality dominated the room. He was a wonderful ambassador for Disney and their new presence in the small town of Orlando.

Ray Fail

Ray Fail was a retired Navy Commander who served as Admiral Fowler's "Chief of Staff". As a former lowly enlisted man, I was always impressed that a high ranking officer like Ray Fail still had to take orders from an Admiral. Everyone has a boss.

Structural and Mechanical Georges

One of Disney's principal consulting engineering firms was **Wheeler and Gray** from California. They had worked with Walt and Admiral Fowler on the design of Disneyland and knew what Disney expected in the way of design and construction standards. Their senior engineer on site was **George Knudsen**, who we called **Structural George** to differentiate him from another George Knudsen, **Mechanical George**, a mechanical engineer who worked for Disney.

Pete Markham

Dames and Moore was Disney's soils consultant, and had also been involved with Disneyland. They knew everything about the soils and what to do to prepare them for foundation construction, roadways, and utilities. **Pete Markham** was an English soils engineer who worked for them. He got promoted and became a manager with Buena Vista Construction Company. Bill Ringo replaced him with Dames and Moore.

Pete usually wore a khaki shirt and khaki pants, and looked a bit military. He was unusual in those days as a construction man who neither smoked nor drank. Most of us did one or the other, usually both. Pete was one of the first guys I ever heard use the term **"construction management"** to define his profession.

Before those days you were either a contractor or an engineer or an architect. An Atlanta firm, **Heery and Heery**, was one of the first to call themselves construction managers. Heery and Heery was the architectural firm that designed what was known originally as Lake Buena Vista Village. It has been expanded over the years and is now known as Disney Springs.

Law Engineering Testing Company, for whom I worked, was in charge of quality control of structural steel fabrication and erection, and concrete production and placement. **Walt Kiser** was in charge early in the project, with me running the steel inspection and testing group, and **John Unterspan** in charge of the concrete technicians. John and I each had senior engineering technicians who supervised the other technicians on a daily basis. My right hand man was **Dick Inge**, and John's right arm was **Howard Hunter.**

Law Engineering Staff on Site

Dick Inge was the lead metals technician. He not only had excellent technical skills, but was a wonderful people person. He was from the area around Lynchburg, Virginia; a little town called Goode, pronounced to rhyme with goo.

Dick had a very gentle nature and was very soft spoken. He was a lay preacher in his church, and a strong Christian with high moral values. I worried about him being rough and tough enough to correct ironworkers when they made errors. An inspector has to have the respect of the inspected, or things can quickly go to hell.

I need not have worried. Dick could do everything the ironworkers could do, even better. They respected his knowledge, and since many of them were born again Christians or card-carrying lay preachers, Dick's ministerial status was a bonus. Dick helped them improve their skills and they treated him like a brother.

I remember Dick as one of the few guys who could walk the steel of the Swiss Family Robinson tree house. The branches were of tubular steel, and it is very hard to walk on a rounded surface, especially way above the ground. Dick would make it look easy, and I was probably one of the few who watched him who knew he had a trick knee that could go out with no notice. He had great balance, and was always prepared for the knee to go.

Howard Hunter was a senior experienced technician who was skilled in concrete, steel, and soil construction techniques. He could always come up with a practical answer to a construction problem. He was a calm man with a commanding presence who had the respect of his technicians and the Disney staff.

Howard was a veteran of World War Two, serving in the Navy as an explosive expert. After the battle of Iwo Jima in the South Pacific he spent days exploding shells on the beach. The waters around the island were filled with the floating bodies of his comrades who had been killed in the fierce . The bodies were covered with crabs feeding on them, and Howard had never been able to eat crabs since that awful experience.

His family was from Flagler County, up near Bunnell in the village of Espanola. His father had been a long time Flagler County commissioner. Howard was a typical Florida cracker. He could identify every tree and plant he saw, and tell you what it meant about the soil it was growing in. He also could remember seeing his first Florida armadillo sometime about 1938. A farmer came roaring into town in his pickup with lights flashing and horn honking. He skidded to a stop in the local gas station and had everyone look in the back of the truck at the strange animal he had just run over outside of town.

Some of the earliest technicians who came aboard in addition to Dick and Howard were Sy Haynes, George Holdridge, Jim Phillips, Jim Quinn, Larry Mondok, Sam Neal, Charlie Folsom, Don Ridley, Charles Hamrick, Larry Carroll, Bobby Geberth, and Ken Wilson. Margie Rushton was our secretary, the glue that held our operation together.

These were some of the best people I ever worked with. I didn't know much about the technical end of this kind of testing and inspection when I started. I learned from my technicians, and I like to think I helped a lot of them with their reporting and management skills.

A Nurseryman and A Truck Counter Get Temporarily Rich

With the tight schedules required to meet the October 1971 opening date, Disney had to trust a lot of people to make things happen. Sometimes that trust was not well placed. Consider the story of the California nurseryman.

The nurseryman was a trusted worker for Disney at Disneyland. They sent him to Orlando in the late 1960s to line up good nurseries and lock in prices for the huge landscaping effort that would take place. The guy allegedly made his own secret deals, buying a couple of nurseries and getting kickbacks from others. It was rumored that he had stashed away more than one million ill-gained dollars before the Disney managers moved to Florida and finally discovered what was going on.

Another Disney employee was employed to count truckloads of fill coming in to various projects at Walt Disney World. Contractors were paid by the truckload. This guy was supposedly on the take, and made a few bucks the easy way. He simply would record phantom trucks that didn't actually deliver fill. He would see that the contractor got paid for those trucks, and take a percentage of the crooked profits under the table from the contractor. Disney managers also finally figured out this guy's game and fired him too.

The Ironworkers

The ironworkers were among my favorite construction workers. Most of them at Walt Disney World were members of **Ironworker Local 808**. Since I was in charge of the people inspecting the work of these guys, I got to know a bit about them and their work culture. Ironworkers are the guys that erect the structural steel framing on buildings. Most of the buildings at Walt Disney World were in this category. The steel framing is usually bolted together in the field, but welding is also typically used.

I inherited a fear of heights from my father. Once when my Dad was working on a job between college terms he was on top of a school bus being built in a factory. His ladder fell down, and he lay spread eagled in terror on the roof of the bus for a long time until a coworker came to rescue him. He refused to climb ladders to put up the storm windows at home – that was my job- and on family trips he would never climb observation towers with the rest of us. Once he even planned a trip from our home in Michigan down to Florida and bypassed the direct route, which would have taken him through the mountains of Tennessee and north Georgia.

Ironworkers spend most of their working time up in the air climbing on structural steel, so I had to confront my fear to be able to work with them. Ironworkers would walk nimbly along a narrow steel beam ten stories above the ground. I would sit on the beam with my feet on the lower flange and shuffle along on my butt in a style the ironworkers referred to as **"coon-assing"**. But they never laughed at me or any other person who was afraid of heights. In fact, they were very supportive and encouraging.

I finally got to where I could fake it and would walk along beams and ride **headache balls** - the heavy ball and hook that hangs from a crane - up to the high stories, but I never got over being scared as hell inside.

When an ironworker fell to his death, the entire brigade of ironworkers took the rest of the day off. There were no questions asked. I can only remember one incident which resulted in death, and that was on the **Contemporary Hotel**. A bundle of sharp edged steel roofing deck was being lifted by a crane to a high story. It came undone and the sheets fanned out and fell on some workers below, killing them.

An ironworker had an assistant, a young guy breaking into the trade, who was known as a **punk**. A foreman was known as a **pusher**. The ironworkers also had their own language. An impact wrench was a **yo-yo**. A vertical steel column was a **col-yume**. They had a sign language to communicate with their brothers on the ground and through the noise of a construction site. If they wanted a 2 inch long ¾ inch diameter bolt tossed up to them, they would hold up two fingers (for two inches), and draw the side of their hand across their neck (3/4 of the way up the body for a ¾ inch bolt), and so on.

Many of the ironworkers were "born again" Christians, and many of them were also lay preachers. Most of them carried what they called "mad money" in their billfolds, usually a hundred dollar bill folded and refolded into the size of a postage stamp. This was hidden from the wife and to be used for fun when things got tight.

One ironworker crew had a pusher who was a karate expert. His favorite thing was going over to the rough bars in Kissimmee after work and picking fights with cowboys. He always won the fights, but a lot of us hoped that one of those cowboys would pull out a six gun and even the odds. The superintendent on this crew was Big Jim, who was 6 foot 5 and weighed a solid 250 pounds to my wimpy 210. The pusher was always trying to get Jim and me into a fight. Fortunately for me, Jim had an even and pleasant disposition, and finally got the pusher to knock off the stuff.

One of my inspectors was very, very picky. It is a hard thing with an inspector to get them to back off. They look at you like you are suggesting they go into the field of prostitution. But this particular inspector was way too picky. It is possible to over inspect any project and flat shut it down. It is done all the time. This is another way of saying that it is impossible to build something exactly the way it is designed and specified. An inspector has to have good judgment as to what is really important and what isn't.

The inspector also had a bit of a mean streak. He loved to find errors in workmanship at the worst possible time, when it created the most inconvenience to the contractor. One day he came screaming into my office and told me that a crane operator had dropped a steel beam from way up on one of the buildings and it had almost hit him. He felt it was not an accident, and demanded I talk to the contractor.

I talked to the man who had been operating the crane and asked him what happened. The guy said he was only trying to warn the inspector to back off, not kill him. He also said it was my word against his. So I moved my guy to another building. I finally found out the best use for him. When a contractor was sloppy and doing bad work, I put this inspector on their project. Finally, after days of super tight inspection, the contractor would beg for another inspector. I would oblige only if they promised to be good and do good work.

The ironworkers were in general good men who were interested in getting their work done and who took a lot of pride in doing it

An Orange County Commissioner Has Issues with Walt Disney World

During those early days of construction, there was at least one man who did not share our enthusiasm for the giant project we were building. That man was Paul Pickett, who at the time was an Orange County Commissioner. Mr. Pickett didn't like the fact that Disney had created its own governmental unit, Reedy Creek Improvement District, and felt like Disney had set things up so as not to pay their fair share of the impacts the new attraction would have on the Orange County taxpayer.

Many times when the commission voted on something in favor of Disney, Pickett was the lone dissenting vote. Looking back through almost 30 years of insight, many people believe Pickett was the only political leader in Central Florida with guts and vision back in those days.

I think Walt Disney would have liked Mr. Pickett in spite of it all.

CHAPTER 11. THE PROJECT TAKES SHAPE ON THE GROUND

Modular Construction At The Contemporary Resort Hotel

Contemporary Hotel Under Construction, Circa 1970

The Contemporary Hotel was built on a site overlooking Bay Lake. The monorails went right through the open ends of the tall lobby. The lobby walls were covered with beautiful murals, and the whole place was quite spectacular for sleepy central Florida. United States Steel Corporation began the construction of the hotel. They had a superintendent on the project that was very unpopular with the workers. They called him **Shiny Shoes** because of the highly polished shoes he wore on the job. Some construction worker even wrote a country song about Shiny Shoes. The song had modest success and was played on the radio quite a bit.

Before the project was completed, Disney took it over and finished construction.

The hotel rooms were manufactured away from the hotel site. They were modular units, designed and built very much like house trailers or mobile homes (what is the difference, does anyone know?). Each room was self-contained, with the bathrooms already installed, the air conditioning units, and everything ready to go except for the furniture. The modules were stacked into the frame of the hotel, plumbing connections made, and the joints between them sealed. It was a revolutionary new way of construction.

The Court of Flags in the Major Realty development at Kirkman Road and I-4 was also built using this modular room technology. The name of this project was changed in later years to Delta Orlando Resort.

United States Steel and Disney tested some of the modular units to make sure they met the fire codes. They burned a few of them, and they went up in flames like tissue paper. There were worried looks on faces all around the test area. Techniques were devised to better fireproof the units before and after they were installed into the hotel frames.

The Excitement of the Walt Disney World Preview Center

Disney built the Preview Center and opened it in early 1970. It was on Lake Buena Vista Boulevard, just off State Road 535. The boulevard has been renamed Hotel Plaza, and is now lined with big hotels.

Lake Buena Vista Boulevard is an elegant tree lined four-lane avenue. It runs into Downtown Disney, which was originally named Lake Buena Vista Village. It was Disney's first major shopping area.

The preview center showed people what was going to be built. I was like most of the people involved in construction of Walt Disney World, taking my wife and kids to see what it was I had been working on all this time. The displays in the Preview Center not only made the site look real to the tourists who would come back down, it also served as an inspiration to those of us working on the project.

This dang thing was real, it would be completed, it would be great, and it would be successful.

The preview center also contained a small post office for the residents and businesses of Lake Buena Vista and Bay Lake. I planned to go into business for myself after Walt Disney World opened, so I rented a box: Post Office Box 25, Lake Buena Vista, Florida.

I printed letterhead and was ready to go. The only problem, in those days, was that nobody had ever heard of Lake Buena Vista, Florida. It was a made up name and a made up town. I was constantly explaining to people that it was out near Disney World, and always had to spell it in those early days.

CHAPTER 12. THE FINAL HECTIC DAYS OF CONSTRUCTION

In the summer of 1971 it was looking to many people like Walt Disney World could never open on its scheduled date, which was October 1, 1971. Schedules had slipped, and there was real concern by Disney executives that there would be enough attractions open to provide a good guest experience. Construction work shifts were lengthened until finally, in the months before the opening, someone was working on the construction 7 days a week 24 hours a day.

The construction people who build one-off projects like Walt Disney World are caught in an inescapable quandary: they have to work themselves out of a job. Unlike a widget factory where the widgets are made year after year in a little factory in Widgetville and a man can spend his life happily making widgets, the construction person lives to build a project and move on to another. Many times there are no other projects to move on to.

Early on in the heat of the project, the construction workers begrudgingly accepted me and my inspectors as necessary evils. They did not like us pointing out defective work and causing them to have to do things over the right way. But as the project neared completion, things changed. Then we were constantly hailed by workers pointing out defective work that we might have overlooked. They were trying to create more work for themselves as the end of the project became more evident.

By September 1971, most of the features of the park were ready to be operated and tested on real people. Before the park was opened to the public, the Disney staff practiced on those of us who had worked on the construction as well as our families. For almost a month we were allowed free entry, free meals, and free rides. The pride we felt was seen in the faces of the guys and gals who had worked long and hard to build the place. It didn't matter to us that we were referred to as human sandbags or test dummies used for fine tuning the rides. We had helped build something special, and it made the long hours and headaches worthwhile.

The excitement was intense in the weeks just before grand opening. Julie Andrews and a dance troupe were in the park rehearsing for a TV special. Michael Jackson and his family group, the Jackson Five, were also around the place for quite a while. Every evening after work the construction workers and their families would get free food and rides and enjoy the new staff smiling and practicing their skills.

On one of these nights I saw the dignified Roy Disney happily sitting on a painted horse bobbing up and down as the Fantasyland Carousel turned merrily round and round. He had a big smile on his face and enjoyed every minute of his ride. Another time I saw him sitting in Dumbo, a ride in which elephants circle around a pylon and weave up and down. Roy believed in trying out everything to make sure it was right for the guests.

Disney Has A Party for the Construction Workers

During these Soft Opening days, Disney threw a party for some of us who had worked on the construction of Walt Disney World. The party was held at a golf cottage owned by Disney at **Bay Hill Country Club**. Most of the brass was there. It was a night for celebration. It was obvious that we would be able to open a wonderful attraction to the public in October 1971. We had met the schedule.

The Disney studio people presented a silent film that showed the construction activity. The photographs had been taken from a tall tower that had been erected on the site in the earliest days of construction. We all knew the camera was up there, gazing down toward **Cinderella's Castle**, but didn't think anything of it. We didn't stop to think how the series of snapshots would freeze what we were doing forever in history.

The film was a series of time-lapse photographs. The camera had automatically taken a photograph every twenty minutes. It was aimed right down at the castle, and you could see the process that had taken many months to accomplish in real time take place before your eyes in several jerky minutes. It was amazing to see the bare ground turn into concrete basement, the basement get covered with buildings, the buildings get covered with fiberglass, trees and shrubs sprouting everywhere. It was hilarious to see you and your friends flitting around the screen like characters in an old Charlie Chaplin silent film.

Construction Workers Look For Permanent Jobs at Walt Disney World

In the weeks before the grand opening, the construction workers and my technicians began to look for permanent jobs at Walt Disney World. Disney did not call them jobs, however. They were "cast positions". A person had to audition to become a member of the Disney cast.

One of my technicians resigned to become a fife player in the Disney fife and drum corps. This group made regular appearances in Liberty Square. I urged him to reconsider, as he was a good technician with a good future with our firm. But he had been sprinkled with the "pixie dust" and would not hear of anything other than his true calling as a Revolutionary War soldier in a colonial uniform tooting a fife.

One entry-level job at Walt Disney World was Herman Pooperscooper. Herman was the name given the Disney employee who followed the horse drawn carriage up and down Main Street. When the horse deposited "horse apples" on the street - and the horse regularly made these deposits - Herman's job was to swoop in quickly and casually pick up the deposit with his long handled scooper.

There was no stigma attached to this job, and no derogatory comments directed toward the Herman of the day on this matter. The most hardened tough as nails construction worker knew this was the ultimate beginning job at Walt Disney World.

If one did a good job in the role of Herman, one could advance upward and onward and become the next **Dick Nunis**, the boss, who started as an hourly worker at Disneyland in 1955 and spent 44 years with the company.

Walt Disney World was a meritocracy. You moved up if you were good at your job and with the guests.

The weeks of the soft opening allowed the workers who would be operating the park to mingle with the workers who had built the park. Those days flowed by until the grand opening seemed anticlimactic.

CHAPTER 13. THE GRAND OPENING

After the month long practice of the soft opening, the Disney staff was ready to greet the paying guests on opening day, October 23, 1971. The relaxed morals of the 1960's were still evident in the early 1970's. One symbol of this new morality and a symbol of the liberated woman was the disappearance of the bra. In the sixties women burned them; in the seventies they simply didn't wear them.

Roy Disney was no prude, but he felt that women without bras were not appropriate for a family park. On opening day he instructed Disney Security and the Main Gate ticket sellers to prohibit any woman who was obviously not wearing a bra from entering the Magic Kingdom.

Shortly after the ticket offices opened, Roy received a panicked call from the Chief of Disney Security. They had turned away so many braless women that the turnstiles were jammed up and there was a huge traffic jam, people wanting to get in who couldn't. It was close to becoming a public relations disaster, and Roy reluctantly withdrew his orders. Those of us who were in the park on opening day can attest that probably half of the women guests that day were not wearing bras. Roy had the right idea at the wrong time.

From the day that it opened it was obvious what a special place Walt Disney World was. I saw a hulk of a biker type that opening day, a Hell's Angels kind of guy with black leather and studs and silver chains and bandana, scraggly beard and tattoos all over his hairy muscular body. He was smoking a cigarette, and rather than flicking his ashes onto the sidewalk he cupped his hand and flicked the ashes into his palm. I watched him finally put the butt and ashes carefully in an ash can.

The clean and wholesome environment brought out the best in the guests and employees. There were plenty of trash bins and ashtrays strategically located throughout the park. Most people used them, and if an errant guest squished out a cigarette on the sidewalk or a kid threw a paper cup on the grass, they were quickly picked up by a courteous Disney worker.

The highlight of the grand opening was seeing Roy Disney give a talk. He was the visible, living link to Walt. He had been there from the beginning to see that Walt Disney World would be what Walt had wanted. He even insisted that the park be renamed from Disney World to Walt Disney World, so that people would never forget who had dreamed up the place. He was one of the many who always asked during thorny construction and creative issues, "What would Walt do?"

Arthur Fiedler of the Boston Pops conducted a large orchestra with musicians assembled from all over the world. The families of Roy and Walt were there. Roy Disney stepped to the microphone and looked at the happy crowd. He thanked all of us who had worked helping to build Walt Disney World.

Then he talked a little bit about Walt:

"My brother Walt and I first went into business together almost a half century ago. And he was really, in my opinion, truly a genius – creative, with great determination, singleness of purpose and drive; and through his entire life he was never pushed off his course or diverted to other things. Walt probably had fewer secrets than any man, because he was always talking to whoever would listen. Talking of story ideas or entertainment projects. My banker one day said, 'How is such and such a project progressing?', and I said, 'Joe, I don't think we have a picture of that name in work.' He repeated the name and said he saw little sketches of the story. I said, 'Joe, Walt was just using you as a good guinea pig to see how you would react to the story. We don't have any picture like that in the work.' And that was the way Walt went through his life."

Roy then talked about Walt's wife, Lillian, who had supported Walt during his entire career. Lillian then joined Roy on the stage. The carillon in Cinderella's Castle was playing **"When You Wish Upon a Star"**.

"Lilly," asked Roy, "you knew all of Walt's ideas and hopes as well as anybody; what would Walt think of it?"

"I think," Lilly replied, "Walt would have approved."

CHAPTER 14. ALL GOOD THINGS MUST COME TO AN END

Walt Disney World was a success from the beginning, but during its construction and after its opening, some problems resulted for its Orlando neighbors.

By the end of 1973, Orlando was in trouble. The Mouse had generated a huge building boom in Orlando, Kissimmee, and surrounding areas. There was suddenly too much of everything. Where in 1969 Horne's was the closest motel, now there were hotels and motels in Lake Buena Vista, sprawling all along US 192 west of Kissimmee, and in Winter Garden. Vacancy rates were the highest in the country, in the 70 per cent range for hotels, motels, and apartments.

Partly built apartment, condominium, and hotel buildings stood abandoned in receivership all around Orlando. Reinforcing bars and structural steel rusted in the Florida sun on building shells that would never be completed.

Clients that I had relied on for my work, like **Gulf Oil Real Estate Development Company** and **Disney Development Company**, literally shut down operations. Other clients like architects and engineers simply closed their doors and left town. Orlando and Florida were gripped in the worst recession since the great depression of the 1930's.

The recession hung on well past the middle 1970's. The economy began to get a bit better in 1976. There was an old joke going around in those days.

Question: "How can you tell if a guy is a former land developer?"

Answer. "He drives around in a 1973 Cadillac Eldorado convertible with a small hole in the trunk lid where his car phone antenna used to be."

Most of us who worked on Walt Disney World moved on to other ventures. It looked to many like Orlando would never recover from its overbuilt situation and recession.

Not many of us were smart enough to look into the future and see the huge metropolis that sleepy little Orlando would become in the following decades.

CHAPTER 15. WHAT WOULD WALT DO IF HE SAW EPCOT TODAY?

What would Walt do?

What would he do if he came back today and saw what has evolved around his Magic Kingdom?

Suppose for a minute that he really is frozen in that apartment up in Cinderella's Castle. Modern medical technology finds a cure for his terminal cancer and Walt is given an injection. Suppose he wakes and rises from his cryogenic chamber and walks over to the windows overlooking the Magic Kingdom.

Disney management would quickly assign a beautiful young female guide to show Walt around. She will hand him a Wall Street Journal, and he will be amazed to learn how much more his Disney stock is worth now than it was back in 1966 when he died.

Walt will quickly get oriented to the Magic Kingdom, because it is pretty much laid out as he created it in his Seventh Preliminary Master Plan. He will get a kick out of the parents and kids riding on the train as the Main Street Railroad circles the park. He will hear the squeals of laughter from Jungle Cruise and the fifes and drums at Liberty Square. He will feel right at home.

He will look out toward Orlando in the east and notice how many tall buildings are there now.

When he last saw it in 1966 it was a typical small Florida town with a few 8 to 10 story buildings dominating its skyline. He'll see the network of new superhighways circling the city and see the traffic jams on nearby Interstate 4. It will remind him of the Santa Ana freeway in Anaheim.

Another look toward the west will reveal Animal Kingdom and the massive development along US-192 west of the Disney main gate.

But Walt looks puzzled, and asks his guide, "Where is Epcot. Where is my City of Tomorrow?"

The guide will direct Walt to look to the south. She points to the dome of Spaceship Earth.

"What is that thing?" asks Walt.

"That is Epcot," replies his guide.

"EPCOT!" roars Walt, "That little dome is EPCOT? Where are the people who are supposed to be living around the city? Where are the shops and businesses that are supposed to be enclosed in the dome?"

"I guess there wasn't enough money to do all that", the guide said.
"Roy could find the money to do that," says Walt. "Why didn't Roy raise the money?"

"Roy got the Magic Kingdom built for you. And he made sure it was done the way you would have done it had you been alive. But he died long before the bean counters and revenue enhancers came up with the concept for Epcot that you see before you. Roy didn't have any input into this Epcot at all."

"But that was the whole idea of building Walt Disney World," said Walt. "That's why I went to all the trouble. Walt Disney World would generate the revenues to build the City of Tomorrow, the Experimental Prototype Community of Tomorrow."

The guide points to a spot along crowded US-192 in the distance to the south.

"Look, Walt," says the guide. It took all of her nerve to call him Walt instead of Mr. Disney.

"Look down toward the highway on the other side of those tee shirt and shell shops. See, over there past that replica of a frontier town, near that shell shop. That's Celebration. It is a planned town of traditional homes and a downtown area that many people feel is close to your original dream of Epcot. There are no slums there, a fine school, a theatre, a church, nice homes, a golf course, and a small hotel."

"People seem to either love Celebration or hate it," said the guide. "The haters compare the people who live in Celebration to the characters in a movie called the Stepford Wives: : a bit artificial, like robots on a stage. But home sales have been

123

wonderful and the stockholders are happy with the way it is going."

As Walt draws his gaze back into his familiar Magic Kingdom, he notices that Herman Pooperscooper is no longer following the horses as they draw the carriage down Main Street. The horses are wearing diapers.

Walt sighs and slowly crawls back into his cryogenic chamber.

"Wake me up when the real Epcot arrives," he tells the guide as he closes the lid.

THE END

EPILOGUE

Mike Miller has lived in Florida since 1960. He graduated from the University of Florida with a degree in civil engineering and has lived and worked in most areas of Florida. His projects include Walt Disney World, EPCOT, Universal Studios and hundreds of commercial, municipal and residential developments all over the state.

What Would Walt Do? is based on Mike's work on the construction of Walt Disney World.

During his career, Mike has developed an understanding and love of Old Florida that is reflected in the pages of his website, **Florida-Backroads-Travel.com,** and a series of travel guides based on the website.

If you find any inaccuracies in this book, please contact Mike at Florida-Backroads-Travel.com and let him know.

If you have enjoyed this book and purchased it at Amazon or any other bookstore, Mike would appreciate it if you would take a couple of minutes to post a short review at Amazon. Thoughtful reviews help other customers make better buying choices. He reads all of his reviews personally, and each one helps him write better books in the future. Thanks for your support!

BOOKS BY MIKE MILLER

Florida Backroads Travel
Northwest Florida Backroads Travel
North Central Florida Backroads Travel
Northeast Florida Backraods Travel
Central East Florida Backroads Travel
Central Florida Backroads Travel
Central West Florida Backroads Travel
Southwest Florida Backroads Travel
Southeast Florida Backroads Travel
Florida Everglades
Florida Wineries
Florida Festivals
Florida Carpenter Gothic Churches
Florida One Tank Trips, Volume 1
Florida Heritage Travel, Vols I, II, III
Living Aboard a Boat
Florida Wineries
What Would Walt Do?